GKSS School of Environmental Research

Series Editors: Hans von Storch • Götz Flöser

Hans von Storch

Richard S.J. Tol

Götz Flöser

Editors

Environmental Crises

Hans von Storch
Richard S.J.Tol
Götz Flöser
(Editors)

Environmental Crises

with 42 Figures

Springer

Professor Hans von Storch
GKSS Forschungszentrum
Inst. Küstenforschung
Max-Planck-Str. 1
21502 Geesthacht
Germany

Professor Dr. Richard S.J.Tol
Economic and Social
Research Institute
Whitaker Square
Sir John Rogerson's Quay
Dublin 2
Ireland

Dr. Götz Flöser
GKSS Forschungszentrum
Inst. Küstenforschung
Max-Planck-Str. 1
21502 Geesthacht
Germany

ISBN 978-3-642-09503-0 e-ISBN 978-3-540-75896-9

Springer is a part of Springer Science+Business Media
springer.com

Cover design: deblik, Berlin

Printed on acid-free paper 30/3180/as 5 4 3 2 1

Contents

Introduction

Hans von Storch and Richard Tol[1]

GKSS Research Centre, Max-Planck-Straße 1, 21502 Geesthacht, Germany

These are the proceedings of the 4th GKSS School on Environmental Research, which took place on 2-11 November 2005 on the Island of Heligoland in the German Bight. 19 PhD students and postdocs from 10 countries came and dared to sail the 3 hours from Cuxhaven to the only rocky island of Germany, the island of Heligoland, in Novembers heavy storm season.

The purpose of this 4th School, with the title "Environmental Crises: Science and Policy" was to study the art and science of analyzing, assessing and anticipating environmental change. Among the issues considered are the observational evidence, statistical analysis and dynamic modeling as well as visioning of not–implausible changes in the environment, the changing public perception of the environment, functions of the environment and its use. A series of four prominent cases, namely climate change, the emissions of gasoline lead into the atmosphere and water bodies, fisheries policies and the management of marine oil pollution is reviewed. Obviously, many other cases could have been chosen, such as Waldsterben, nuclear energy, release of carcinogenic substances, genetically modified food, threat to biodiversity, import of alien species, just to mention a few. We had chosen cases which were close to the research interests of the organizers Hans von Storch and Richard Tol and their institutions, namely the Institute for Coastal Research of the GKSS Research Center in Geesthacht and the Centre for Marine and Atmospheric Sciences (ZMAW) [1] in Hamburg.

The first three Schools have produced already three proceedings, namely:

[1] Hans von Storch is director of the Institute for Coastal Research, and professor at the Meteorological Institute of the University of Hamburg, which is part of the ZMAW. Richard Tol was professor at the Research Unit Sustainability and Global Change of the University of Hamburg, which is also part of ZMAW. In the meantime, Professor Tol has moved to Economic and Social Research Institute, (Dublin, Ireland). He holds also professorships at the Institute for Environmental Studies (Vrije Universiteit, Amsterdam, The Netherlands) and at Carnegie Mellon University (Engineering and Public Policy, Pittsburgh, PA, USA)

1. von Storch, H., and G. Flöser (Eds.), 1999: *Anthropogenic Climate Change.* Proceedings of the First GKSS School on Environmental Research, Springer Verlag, ISBN 3-540-65033-4
2. von Storch, H., and G. Flöser (Eds.), 2001: *Models in Environmental Research.* Proceedings of the Second GKSS School on Environmental Research, Springer Verlag ISBN 3-540
3. Fischer, H., T. Kumke, G. Lohmann, G. Flöser, H. Miller, H von Storch und J. F. W. Negendank, 2004 (Eds.): *The Climate in Historical Times. Towards a Synthesis of Holocene Proxy Data and Climate Models.* Proceedings of the Third GKSS School on Environmental Research, Springer Verlag, Berlin - Heidelberg - New York, 487 pp., ISBN 3-540-20601-9

In the first two of these Schools, both natural and social sciences have already been considered. The background of this combination of natural and social issues, which was again at the center of this School, is that environmental sciences are in many if not most cases subject to public discourses in the media, and that they provide input to the political decision processes. Environmental research is hardly independent from public needs, concerns and visions, perceptions and culturally constructed explanations and even politics. Thus, when engaging in environmental research, scientists must understand both the needs and expectations on the public side; the mechanisms, which relay and transform the scientific knowledge into the public arena; the commitment of scientists to certain value bases, which may influence his/her analytical functioning. In the four cases of this volume, these issues are addressed.

Thus, the main message of this School, and these proceedings is: natural science analysis of socially significant issue needs to be accompanied by economic or social analysis, when providing useful input into the public and policy arena. The link to social sciences is also important for helping the natural scientist to respect his limits of competence; to deal appropriately with media; to avoid the often seen normative driven, nave practice of prescriptively "advising" the public and stakeholders[2].

Also this School and this book have come into being only with the help of many other persons in particular Sönke Rau and Mrs. Liesner. Hans von Storch, 31. Juli 2007

[2] See also Pielke, R.A., jr., 2007: The Honest Broker: Making Sense of Science in Policy and Politics. Cambridge University Press.

Climate Change Scenarios – Purpose and Construction

Hans von Storch

GKSS Research Centre, Max-Planck-Straße 1, 21502 Geesthacht, Germany

1 Introduction

Scenarios are descriptions of possible futures – of different plausible futures. Scenarios are not predictions but "storyboards", a series of alternative visions of futures, which are possible, plausible, internally consistent but not necessarily probable (e.g. Schwartz, 1991; see also Tol, 2007b). The purpose of scenarios is to confront stakeholders and policymakers with possible future conditions so that they can analyze the availability and usefulness of options to confront the unknown future. Scenarios allow implementing measures now to avoid unwanted futures; they may also be used to increase the chances for the emergence of favorable futures.

In daily life, we are operating frequently with scenarios. For instance, when planning in spring for a children's birthday party next summer, we consider the scenarios of an outdoor party on a sunny day or an indoor party on a rainy day. Both scenarios are possible, plausible and internally consistent. Planning for a snowy day, on the other hand, is not considered, as this would be an inconsistent scenario.

In climate research, scenarios have been widely used since the introduction of the IPCC process at the end of the 1980s (Houghton et al., 1990, 1992, 1996, 2001). These scenarios are built in a series of steps. This series begins with scenarios of emissions of radiatively active substances, i.e., greenhouse gases, such as carbon dioxide or methane, and aerosols. These scenarios depend on a variety of developments unrelated to climate itself, in particular on population growth, efficiency of energy use, and technological development (Tol, 2007b). Many of these factors are unpredictable; therefore, a variety of sometimes ad-hoc assumptions are entering these scenarios.

In the next steps, the construction of scenarios is less ad-hoc, as they essentially process the emissions scenarios. The first step is to transform the emissions into atmospheric concentrations, which are then fed into global climate models[1]. Thus, from possible, plausible and internally consistent future emissions are derived estimates of possible, plausible and internally consistent future climate, i.e., seasonal means, ranges of variability, spectra, or spatial patterns. These are the global climate change scenarios. Effects of changes in the natural climate forcing factors such as changes in sun radiation, or volcano eruptions are often not considered in these scenarios. Instead the scenarios try to vision what will happen in the future depending on anthropogenic changes (even if changes of land-use are usually not considered). We may thus call them *anthropogenic climate change scenarios.*

The global climate models are supposed to describe climate dynamics on spatial scales[2] of, say, 1000 km and more. They do not resolve the geographical features such as the details of the Baltic Sea catchment. For instance, in the global models, the Baltic Sea is not connected to the North Sea through narrow sills; instead, the Baltic Sea is something like an extension of the North Sea with a broad link. In addition, the Scandinavian mountain range

[1] There are different types of climate models (cf. McGuffie and Henderson-Sellers, 1997; Crowley and North, 1991; von Storch and Flöser, 2001 or Müller and von Storch, 2005). The simplest form are energy balance models, which describe in a rather schematic way the flux and fate of energy entering the atmosphere as solar radiation and leaving it as long- and short wave radiation. These models are meant as conceptual, minimum complexity tools, to describe the fundamental aspects o the thermodynamic engine "climate system". At the other end of complexity are the maximum complexity tools, which contain as many processes and details as can be processed on a contemporary computer. Being limited by the computational resources, these models grow in complexity over time – simply because the computers become continuously more powerful. Such models are supposed to approximate the complexity of the real system. They simulate a sequence of hourly, or even more frequently sampled, weather, with very many atmospheric, oceanic and cryospheric variables – such as temperature, salinity, wind speed, cloud water content, upwelling, ice thickness etc. From these multiple time series, the required statistics (= climate) are derived. Thus, working with simulated data is similar to working with observed data. The only, and significant, difference is that one can perform experiments with climate models, which is impossible with the real world. However, present day climate models are coarse and the modelling of several processes related to the water and energy cycles are not well understood and only crudely described in these models.

[2] "Scales" is a fundamental concept in climate science. The term refers to typical lengths or typical durations of phenomena or processes. Obviously, this definition is fuzzy. Scales necessarily refer only to orders of magnitude. Global scales refer to several thousand kilometres and more; the continental scale to a few thousand kilometres and more, and regional scales to hundred kilometres and more. For instance, the Baltic Sea is a regional feature of the global climate system.

When constructing climate models the equations can only be resolved within a limited resolution. Dynamic features larger than the grid domain need to be prescribed, while features below the grid size need to be parameterized. Typical atmospheric and oceanic processes that need to be parameterized in climate models are indicated in Fig. 2.

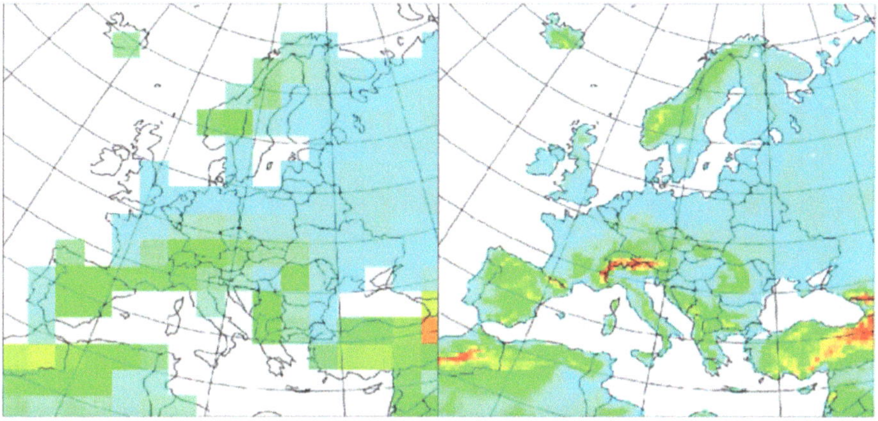

Fig. 1. Typical different atmospheric model grid resolutions with corresponding land masks. *Left:* T42 used in global models, *right:* 50 km grid used in regional models (courtesy: Ole Bøssing-Christensen)

is shallow (Fig. 1). Therefore, in a second step, possible future changes of regional scale climatic features are derived using regional climate models[3]. These models are based on the concept of "downscaling", according to which smaller scale weather statistics (regional climate) are the result of a dynamical interplay of larger-scale weather (continental and global climate) and regional physiographic detail (von Storch, 1995, 1999). There are two approaches for downscaling (see e.g. Giorgi et al., 2001) – namely empirical or statistical downscaling, which employs statistically fitted links between variables representative for the large-scale weather state or weather statistics, and locally or regionally significant variables. The alternative is dynamical downscaling, which employs a regional climate model. Such regional models are constrained by the large-scale state simulated by the global models along the lateral boundaries and sometimes in the interior. With horizontal grid sizes

[3] Regional climate models are built in the same way as global climate models – with the only and significant difference that they are set up on a limited domain with time-variable lateral boundary conditions. Mathematically this is not a well-posed problem, i.e., there is not always one and only one solution satisfying both the ruling differential equations and the boundary constraints. By including a lateral "sponge-zone" along the lateral boundaries, within which the internal solution and the externally given boundary conditions are nudged, it is practically ensured that there is a solution, and that instabilities are avoided. Most present day regional climate models only downscale the global models and they thus do not send the information back to the global scale. This implies they are strongly controlled by the global climate model.

The advantage of regional climate models is that they provide an increased horizontal resolution. Therefore, they can simulate regional detail as is required for many impact studies.

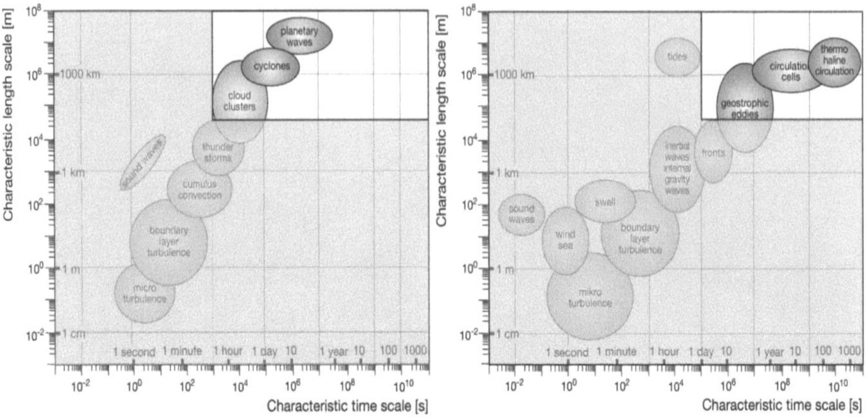

Fig. 2. Spatial and temporal scales of some major processes in the atmosphere (**a**) and in the ocean (**b**). In the figures the grey area represents sub-grid scales in common global models, while the visible parts in the upper right corners represent the processes which climate models often aim to resolve (redrawn from Müller and von Storch 2004)

of typically 10 to 50 km, such models resolve features with minimum scales of some tens to a few hundred of kilometers. They also simulate the emergence of rare events, such as strong rainfall episodes and strong windstorms.

Methodologically, the anthropogenic climate change scenarios are conditional predictions. After the emissions scenarios are given, no further ad-hoc decisions are required. Significant assumptions are required only for the design of the emission scenarios. These assumptions refer to socio-economic processes, which lead to emissions. Each set of socio-economic assumptions is associated with another climate development. However, they all share a number of changes, namely an increase in air and sea surface temperature and a rise in water levels. Thus, the conditional predictions agree independently of the specific conditioning assumptions on a general increase in temperature and sea level – and these conditional predictions become unconditional predictions with respect to these properties.

When dealing with regional or local scenarios, one has to keep in mind that not only global and regional climate changes, but that the effect of climate on regional and local economic and ecological conditions depends on many other factors, which may also change. For instance, changing land-use may locally be of comparable significance as the influence of climate changes. Thus, a complete analysis on the regional level needs an additional set of scenarios which describes the changing usage of the local and regional environment (e.g. Grossmann, 2005, 2006; Bray et al., 2003).

2 Emission Scenarios

A number of emission scenarios has been published as "IPCC Special Report on Emissions Scenarios" (SRES; http://www.grida.no/climate/ipcc/emission) prepared by economists and other social scientists for the Third Assessment report of the IPCC (see also Tol, 2007b). They utilize scenarios of greenhouse gas and aerosol emissions, or of changing land use.

(A1) a world of rapid economic growth and rapid introduction of new and more efficient technology,
(A2) a very heterogeneous world with an emphasis on family values and local traditions,
(B1) a world of "dematerialization" and introduction of clean technologies,
(B2) a world with an emphasis on local solutions to economic and environmental sustainability.

The scenarios do not anticipate any specific mitigation policies for avoiding climate change. The authors emphasize "no explicit judgments have been made by the SRES team as to their desirability or probability".

The Scenarios A2 and B2 were used widely in recent years. Therefore, we explain the socio-economic background of these scenarios in more detail (for a summary of the other two scenarios, refer to Müller and von Storch, 2004):

SRES describes the A2-scenario as follows: "... characterized by lower trade flows, relatively slow capital stock turnover, and slower technological change. The world "consolidates" into a series of economic regions. Self-reliance in terms of resources and less emphasis on economic, social, and cultural interactions between regions are characteristic for this future. Economic growth is uneven and the income gap between now-industrialized and developing parts of the world does not narrow. People, ideas, and capital are less mobile so that technology diffuses more slowly. International disparities in productivity, and hence income per capita, are largely maintained or increased in absolute terms. With the emphasis on family and community life, fertility rates decline relatively slowly, which makes the population the largest among the storylines (15 billion by 2100). Technological change is more heterogeneous. Regions with abundant energy and mineral resources evolve more resource-intensive economies, while those poor in resources place a very high priority on minimizing import dependence through technological innovation to improve resource efficiency and make use of substitute inputs. Energy use per unit of GDP declines with a pace of 0.5 to 0.7% per year. Social and political structures diversify; some regions move toward stronger welfare systems and reduced income inequality, while others move toward "leaner" government and more heterogeneous income distributions. With substantial food requirements, agricultural productivity is one of the main focus areas for innovation and research, development efforts and environmental concerns. Global environmental concerns are relatively weak."

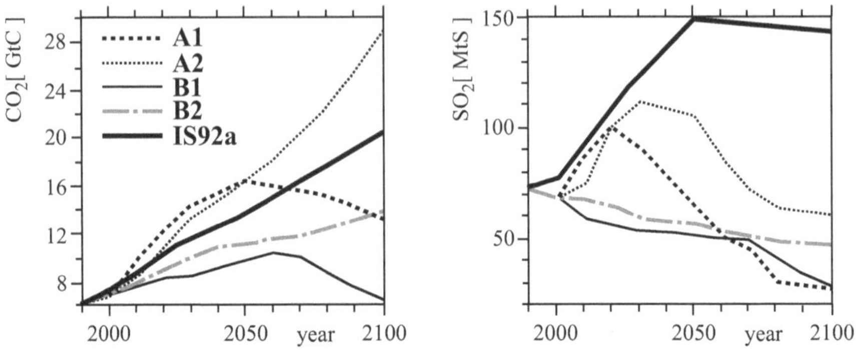

Fig. 3. Scenarios of possible, plausible, internally consistent but not necessarily probable future emissions of carbon dioxide (a representative of greenhouse gases; in gigatons C) and of sulphate dioxide (a representative of anthropogenic aerosols; in megatons S). A1, B1, A2 and B2 are provided by SRES, IS92a is a scenario used in the Second Assessment Report of the IPCC in 1995

In B2, there is "... increased concern for environmental and social sustainability. Increasingly, government policies and business strategies at the national and local levels are influenced by environmentally aware citizens, with a trend toward local self-reliance and stronger communities. Human welfare, equality, and environmental protection all have high priority, and they are addressed through community-based social solutions in addition to technical solutions. Education and welfare programs are pursued widely, which reduces mortality and fertility. The population reaches about 10 billion people by 2100. Income per capita grows at an intermediate rate. The high educational levels promote both development and environmental protection. Environmental protection is one of the few truly international common priorities. However, strategies to address global environmental challenges are not of a central priority and are thus less successful compared to local and regional environmental response strategies. The governments have difficulty designing and implementing agreements that combine global environmental protection. Land-use management becomes better integrated at the local level. Urban and transport infrastructure is a particular focus of community innovation, and contributes to a low level of car dependence and less urban sprawl. An emphasis on food self-reliance contributes to a shift in dietary patterns toward local produce, with relatively low meat consumption in countries with high population densities. Energy systems differ from region to region. The need to use energy and other resources more efficiently spurs the development of less carbon-intensive technology in some regions. Although globally the energy system remains predominantly hydrocarbon-based, a gradual transition occurs away from the current share of fossil resources in world energy supply."

Expected emissions of greenhouse gases and aerosols into the atmosphere are derived from these assumptions and descriptions. Figure 3 shows the expected SRES scenarios for carbon dioxide (a representative of greenhouse gases; in gigatons per year) and sulfate dioxide (a representative of anthropogenic aerosols; in megatons sulfur).

The SRES scenarios are not unanimously accepted by the economic community. Some researchers find the scenarios internally inconsistent (Tol, 2007b). A documentation of the various points raised is provided by the Select Committee of Economic Affairs of the House of Lords in London (2005). A key critique is that the expectation of economic growth in different parts of the world is based on market exchange ranges (MER) and not on purchasing power parity (PPP). Another aspect is the implicit assumption in the SRES scenarios that the difference in income between developing and developed countries will significantly shrink until the end of this century (Tol, 2007a, 2007b). These assumptions, the argument is, lead to an exaggeration of expected future emissions.

3 Scenarios of Anthropogenic Global Climate Change

The emission scenarios are first transformed into scenarios of atmospheric loadings of greenhouse gases and aerosols. Then, the global climate models derive from these concentrations – without any other externally set specifications – sequences of hourly weather for typically a hundred years. A large number of relevant variables are calculated, for the troposphere, the lower stratosphere and the oceans, but also at the different boundaries of land, air, ocean and sea ice – such as air temperature, soil temperature, sea surface temperature, precipitation, salinity, sea ice coverage or wind speed.

Global climate models suffer to some degree from systematic error, so-called biases. Globally, these errors are not large, but they can be regionally large and are too large to determine the expected climate changes only from a simulation with, for instance, elevated greenhouse gases concentrations. Instead, the climate change is determined by comparing the statistics of a "scenario simulation" with anthropogenic greenhouse and aerosol levels with the statistics of a "control run", which is supposed to represent contemporary conditions with unchanged atmospheric composition. The difference between the control run and present climate conditions provides us with a measure of the quality of the climate simulation. If this difference is large compared to the scenario change, the climate simulation should be interpreted with care.

Figure 4 shows, as an example, the expected change of winter air temperature for the last 30 years of the 21st century in the scenario A2 and in the scenario B2. This change is given as the difference of 30 years mean in the scenario run and in the control run. The air temperature rises almost everywhere; the increase is larger in the higher-concentration scenario A2 than in the lower-concentration B2. Temperatures over land rise faster compared to those over the oceans, which are thermally more inert than air temperatures

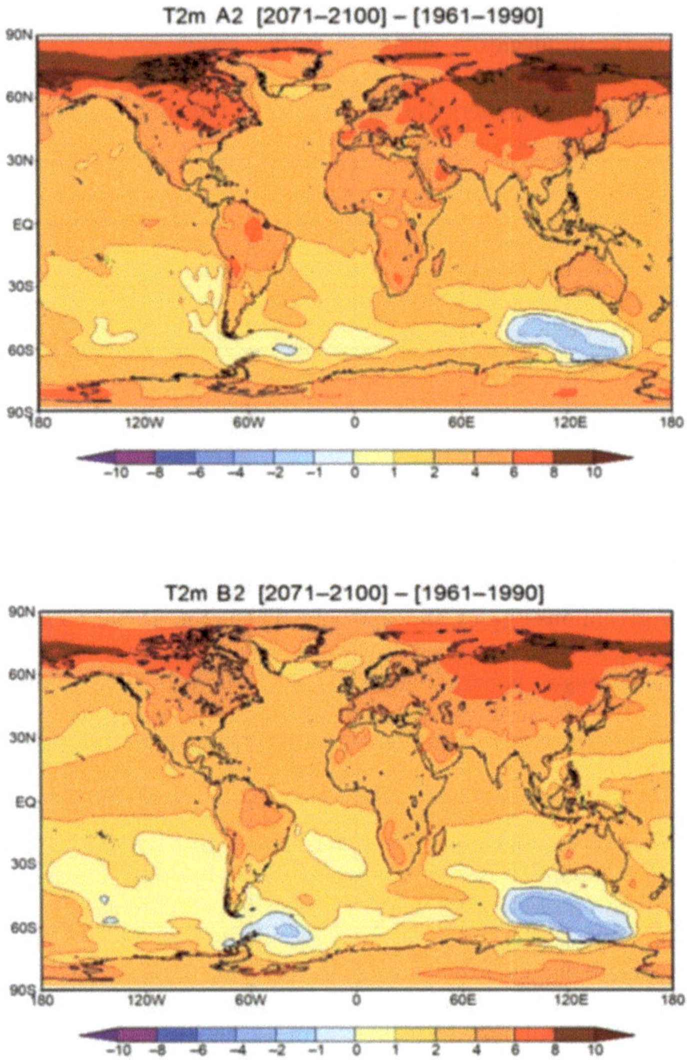

Fig. 4. Global scenarios of air temperature change (in K) at the end of the 21st century, as determined by a global climate model forced with A2 and B2 emissions. Courtesy: Danmarks Meteorologiske Institut

over land. In Arctic regions, the increase is particularly strong and this is related to the partial melting of permafrost and sea ice.

4 Regional Anthropogenic Climate Change

A number of projects, as e.g. PRUDENCE (Christensen et al., 2002), have used regional climate models to derive regional climate change scenarios for Central and Northern Europe. A major result was that the regional models return rather similar scenarios when they are forced by the same global climate change scenario (e.g., Woth et al., 2005; Déqué et al., 2005) – so that the choice of the regional climate model is of minor relevance.

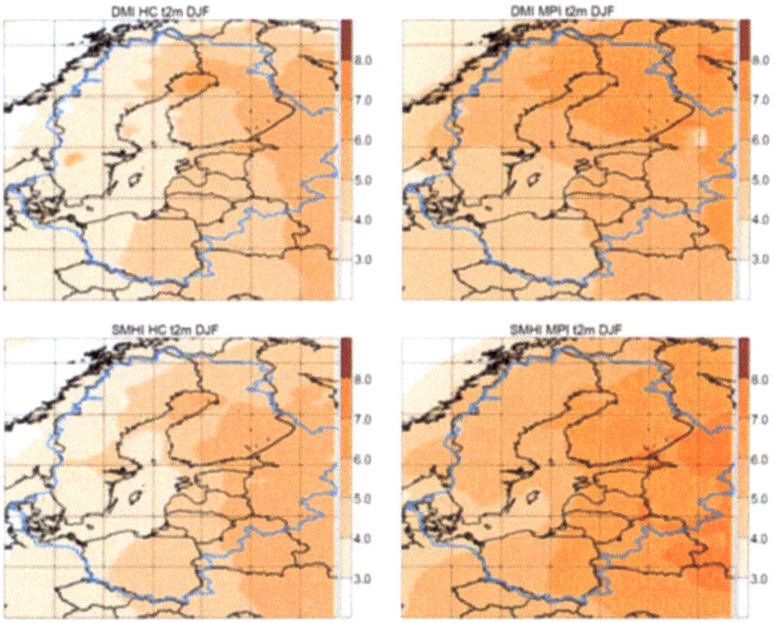

Fig. 5. Temperature change (in K) for winter (DJF) between the periods 1961–1990 and 2071–2100 according to the SRES A2 scenario. Plots on the left used HadAM3H as boundary conditions, plots on the right used ECHAM4/OPYC3 as boundaries. For each season the upper row is the DMI regional model HIRHAM, the lower row is the SMHI model RCAO. Note that the ECHAM4/OPYC3 scenario simulations used as boundaries are different for the two downscaling experiments. The Baltic Sea catchment is indicated by the thick pale blue contour

However, if different driving global climate change scenarios are used – by using different emission scenarios or different global climate models – the differences become larger (e.g., Woth, 2005; Déqué et al., 2005). If higher levels of anthropogenic forcing are applied, then the regional changes become stronger on average, even if not necessarily in a statistically significant manner. This lack of significance is related to the fact that the signal-to-noise ratio of systematic change and weather noise gets smaller if the considered spatial scales are reduced.

A joint feature of all regional simulations is that it is getting warmer; frequency distributions tend to become broader with respect to central European summer rainfall (Christensen and Christensen, 2003) and North Sea winter wind speeds (Woth, 2005).

As an example, the expected changes of winter (DJF – December/January/February) mean temperatures are shown in Fig. 5. The different model configurations (two global models, two regional models) indicate that when the snow cover retreats to the north and to the east, the climate in the Baltic Sea catchment undergoes large changes. A common feature in all regional downscaling experiments is the stronger increase in wintertime temperatures compared to summertime temperatures in the northern and eastern part of the Baltic Sea catchment (e.g., Christensen et al., 2003; Déqué et al., 2005).

Literature

Bray D, Hagner C, Grossmann I (2003) Grey, Green, Big Blue: three regional development scenarios addressing the future of Schleswig–Holstein. GKSS–Report 2003/25, GKSS Research Center, Geesthacht 2003

Christensen JH, Christensen OB (2003) Severe summertime flooding in Europe. Nature 421:805–806

Christensen JH, Carter T, Giorgi F (2002) PRUDENCE employs new methods to assess European climate change. EOS 83:147

Crowley TJ, North GR (1991) Paleoclimatology. Oxford University Press, New York, 330 pp

Déqué M, Jones RG, Wild M, Giorgi F, Christensen JH, Hassell DC, Vidale PL, Rockel B, Jacob D, Kjellström E, de Castro M, Kucharski F, van den Hurk B (2005) Global high resolution versus Limited Area Model climate change projections over Europe: quantifying confidence level from PRUDENCE results. Clim Dyn 25,6:653670. doi: 10.1007/s00382-005-0052-1

Giorgi F, Hewitson B, Christensen J, Hulme M, von Storch H, Whetton P, Jones R, Mearns L, Fu C (2001) Regional climate information – evaluation and projections. In: Houghton JT et al (eds) Climate Change 2001. The Scientific Basis, Cambridge University Press, 583–638

Grossmann I (2005) Future Perspectives for the Lower Elbe Region 2000–2030: Climate Trends and Globalisation. PhD Thesis, Hamburg University

Grossmann I (2006) Three scenarios of the greater Hamburg region. Futures 38,1:31–49

Houghton JT, Callander BA, Varney SK (eds) (1992) Climate Change 1992. Cambridge University Press, 200pp

Houghton JT, Ding Y, Griggs DJ Noguer M, van der Linden PJ, Dai X, Maskell K, Johnson CA (2001) Climate Change 2001: The Scientific Basis. Cambridge University Press, 881pp

Houghton JL, Jenkins GJ, Ephraums JJ (eds) (1990) Climate Change. The IPCC scientific assessment. Cambridge University Press, 365pp

Houghton JT, Meira Filho LG, Callander BA, Harris N, Kattenberg A, Maskell K (eds) (1996) Climate Change 1995. The Science of Climate Change. Cambridge University Press, 572 pp

House of Lords, Select Committee on Economic Affairs (2005) The Economics of Climate Change. Volume I: Report, 2nd Report of Session 2005-06, Authority of the House of Lords, London, UK; The Stationery Office Limited, HL Paper 12-I (http://www.publications.parliament.uk/pa/ld/ldeconaf.htm# evid)

McGuffie K, Henderson-Sellers A (1997) A climate modelling primer. 2nd edition, John Wiley & Sons, Chichester, Great Britain, 253pp

Müller P, von Storch H (2004) Computer Modelling in Atmospheric and Oceanic Sciences: Building Knowledge. Springer, Berlin Heidelberg New York, 304pp

Schwartz P (1991) The art of the long view. John Wiley & Sons, 272pp

Tol RSJ (2007a) Exchange rates and climate change: An application of FUND. Climatic Change 75: 59–80, doi: 10.1007/s10584-005-9003-4

Tol RSJ (2007b) Economic scenarios for global change, this volume

von Storch H (1995) Inconsistencies at the interface of climate impact studies and global climate research. Meteorol Zeitschrift 4 NF:72–80

von Storch H (1999) The global and regional climate system. In: von Storch H, Flöser G (eds) Anthropogenic Climate Change, Springer

von Storch H, Flöser G (eds) (2001) Models in Environmental Research. Proceedings of the Second GKSS School on Environmental Research, Springer

Woth K, Weisse R, von Storch H (2005) Dynamical modelling of North Sea storm surge extremes under climate change conditions – an ensemble study. Ocean Dyn. DOI 10.1007/s10236-005-0024-3

Woth K (2005) Projections of North Sea storm surge extremes in a warmer climate: How important are the RCM driving GCM and the chosen scenario? Geophys Res Lett 32, L22708, doi: 10.1029/2005GL023762

Economic Scenarios for Global Change

Richard S.J. Tol[1,2,3]

[1] Economic and Social Research Institute, Dublin, Ireland
[2] Institute for Environmental Studies, Vrije Universiteit, Amsterdam, The Netherlands
[3] Engineering and Public Policy, Carnegie Mellon University, Pittsburgh, PA, USA

1 Introduction

Scenarios are almost invariably used in the analysis and assessment of environmental change, particularly if the aim is to inform policy or support decision-making. Scenarios often drive the conclusions of such studies. It is therefore important to understand what scenarios are; to be able to distinguish between the good and the bad; and, from this understanding, to improve the scenarios. This chapter contributes to that, with a particular focus on economic scenarios. The focus is on understanding rather than technical detail.

In Sect. 2, I define scenarios, their purposes, and economic scenarios. In Sect. 3, I discuss predictability, scale, and resolution – issues that are closely related. In Sect. 4, I turn to some of the practical issues of scenario building. Sections 5, 6, and 7 include discussion scenarios of economic growth, greenhouse gas emissions, and vulnerability to climate change, respectively. Section 8 treats the empirical validity of scenarios, using IPCC SRES as an example. Section 9 forms the conclusions.

2 What are Economic Scenarios?

Scenarios are an important element of assessment and decision support. Assessment and decision support are about the future, or rather about the decisions we need to make today or tomorrow to make the future more desirable. Scenarios describe alternative futures. Decisions are favourable if we can reasonably expect that the future would be brighter as a result, or if at least we have good reason to expect that the future would not be worse. Scenarios provide the basis for these judgements (Yohe, 1991).

Some people elevate scenarios to a separate level (van Notten et al., 2005; Postma and Liebl, 2005). I do not think this is necessary; and it may even be counterproductive. Scenarios are not forecasts. Forecasts are predictions about the future. "The sun will rise tomorrow at 6:42 am" is an example of a forecast. Scenarios are not conditional forecasts either. "The global mean temperature would rise if the atmospheric concentration of carbon dioxide continues to increase" is an example of a conditional forecast. The crucial distinction between a forecast and conditional forecast is that the latter is predicated on developments outside the domain of the forecast. If one is willing to predict that atmospheric CO_2 concentrations will rise, then "the temperature will rise" becomes a forecast.

Most people would forecast that carbon dioxide concentrations will rise (Nakicenovic et al., 1998). Forecasts are informative only if they are specific. The following statement: "The atmospheric concentration of carbon dioxide will exceed 700 ppm in 2100 if oil and gas prices remain high, if fear of proliferation prevents a nuclear renaissance, and if climate policy is weak" looks like a conditional forecast – and in a mathematical sense, it is. Most people, however, would call it a scenario.

The distinction I make here between a conditional forecast and a scenario is that the former is limited to well-understood processes (the forecast) driven by well-defined boundary conditions (the conditions). In contrast, scenarios also contain elements which are not well enough understood to model, both in the forecast and its boundary conditions. In the example above, the oil and gas market are not very predictable, and the public mood on nuclear power is even less so.

Scenarios have been defined as "internally consistent descriptions of alternative futures" (Ericksen, 1975). However, as scenarios contain pre-methodological elements,[4] internal consistency cannot be guaranteed. Only a mathematical model can guarantee internal consistency.[5] I therefore prefer to define scenarios as "not-implausible descriptions of alternative futures" (Yohe et al., 1999; Strzepek et al., 2001).

Scenarios are special because they are a weak form of a conditional forecast. Research should focus on making scenarios less special, that is, more like conditional forecasts. Attempts to keep scenarios special, in fact are attempts to weaken the field.

Scenarios are often interpreted as (unconditional) forecasts. The temperature scenarios of the IPCC TAR range from 1°C to 6°C warming by 2100 (IPCC 2001). These results depend on assumptions on future fertility in Europe, the oil reserves of Saudi Arabia, and acidification policy in China (Nakicenovic and Swart, 2001). These processes are not understood, as the people who build the emissions scenarios would readily admit. Nonetheless, the IPCC scenarios are often presented as forecasts – this mistake is commonly made by journalists and policy makers, but scientists are not without fault either.

[4] In its pre-methodological stage, a field of research is characterised by a lack of rigorous methods, so that results are not replicable and depend on the analyst as much as on the analysis.

[5] Note that there are also internally inconsistent mathematical models, also in the supposedly peer-reviewed literature.

The proper use of scenarios is in exploration of the future – What could happen? What may the future hold? – and in assessment of policy – What could happen if this law were implemented? How can undesirable developments be avoided? Rather than finding the optimal policy for a particular future, scenarios are best used as tests for the robustness of policy strategies. If a particular policy performs well under specific circumstances, but badly under other, just as plausible conditions, then perhaps it should be replaced by a policy that does reasonably well on all occasions (Lempert et al., 2004). Therefore, one should not build a scenario as a "not-implausible description of the future". Rather, one should build scenarios as "not-implausible descriptions of alternative futures". The plurality of scenarios is crucial.

Scenarios are often used as a warning: "if things continue like they have, bad stuff would happen". This is an extension of the above. In this case, scenarios are used to demonstrate the lack of robustness of current policies. Note that, in this case, the scenario is the opposite of a forecast. The scenario is, or is supposed to be, a self-defying prophecy. The "forecast" is issued so that it may not come true.

Economic scenarios are like all other scenarios, but they are about economies and economic variables. One can interpret this in two ways. In the broad interpretation, economic scenarios are about everything that is researched in the academic discipline of economics. This is broad indeed, as economics has expanded into crime, ethics, faith, geography, health, marriage, politics, and war. In fact, economics is now best seen as the application of mathematical techniques to the study of human behaviour. Some economists are pushing to extend this to animal (Glimcher, 2003) and even plant behaviour (Tschirhart, 2000).

I will use a narrower interpretation. Here, economic scenarios are about the economy, that is, such things as production and consumption, prices and incomes, employment and investment.

3 Predictability, Resolution and Scale

Some economic variables are very predictable. People go to work in the morning, they go shopping on Saturday, and on holiday in August. These things change very slowly, if at all.[6]

Because people and companies are set in their ways and decisions take time to materialise into facts on the ground, one can predict many economic

[6] In a peculiar semantic twist, geoscientists reserve the word "predictability" for the ability to predict anomalies, that is, deviations from normalcy. This of course requires one to assume that the norm has not changed, which is in itself a prediction. At the same time, predicting trivial things (it will be warmer in August than in January) does not require a lot of skill (apart from knowing roughly where one is). In finance, the litmus test for predictability is that the model systematically outperforms the nave forecast based on the efficient market hypothesis.

variables with a fair degree of confidence for 12 to 18 months into the future. After that, predictive skill fades rapidly. This is because expectations matter. If consumers are confident about their future income, and businesspeople confident about future sales, money will flow. People will book holidays, buy houses and cars. Companies will build new plants and hire new workers. When the mood changes, many such decisions are put on hold. People's moods are very hard to predict, as they depend on external factors such as terrorism and the success of a nation's soccer team, and on internal factors such as hype and hysteria. The mood may change overnight, but it takes a year or so to have noticeable effect on the economy. The same is true for such things as oil shocks and the introduction of unexpected technologies. This is because many of the substantial things that happen today were in fact decided some time ago (Samuelson et al., 2001; Romer, 1996).

This lack of predictability should not be overstated. It is hard to predict whether a developed economic will grow by 1 or 3% in two years, but 1% or +5% are hard to conceive. Emerging and under-developing economies are more volatile, but this volatility is still within bounds – apart from war and political meltdown, which can put economies into a free fall.

Although economic variability is hard to predict, economic growth is easier. As growth is the dominant process in the longer term, economies become more predictable again for a time horizon of ten years and longer. This is because moods may be fickle, but people are not delusional. The average person is sometimes overly pessimistic, and sometimes overly optimistic – but on average, she is realistic – evolution would not tolerate anything else (Weibull, 1995).

In the longer term, predictions are conditional first and foremost on policy. Extending the time horizon even further, technological change becomes the main driver of economic growth. Policy loses its prime position to technology, because, just as irrational expectations are impossible to maintain against the observations, bad policies will not survive repeated elections and international competition. Technology is a cumulative process, however. This does not only lead to exponential growth of the economy, but also to exponential growth of the uncertainties (Barro and Sala-i-Martin, 1995; Romer, 1996).

The spatial resolution and scope of scenarios is important too. Economic statistics are primarily collected at national level. Only some data are available at a finer resolution. This implies that most economic models and most economic scenarios concern countries and aggregates of countries. There is rapid progress in new geographical economics, with regard to both theory and data, but this body of work is not yet ready for application (Brakman et al., 2001; Fujita et al., 1999). Spatially explicit scenarios of economic variables are not to be trusted, as they lack a solid theoretical and empirical foundation.

Economic scenarios for a city, an island or a region should also be interpreted with caution. Although it may be possible to collect substantial data for a confined location, the crucial decisions are often made outside the place of interest. For instance, research is an important component of the Helgoland economy. Funding decisions on Helgoland's research facilities and staff are made in Bremerhaven, Bonn, and Brussels. This implies that the boundary conditions dominate the analysis, and that the results are a trivial mapping of the assumptions. Scenario building degenerates to a tautological process, of the kind "if we assume rapid economic growth, people will be much richer in the future".

Economic data and models are better at resolving sectoral detail than spatial detail. If necessary, considerable technological detail can be added (Boehringer and Loeschl, 2003). Typically, this adds process knowledge to the model and enhances the realism and robustness of the results. In the longer term, problems may arise, however. First, the sectoral classification of economic data reflects the importance of economic sectors at the time the classification was made. Tourism and recreation were trivial economic activities in 1950 when the foundations were laid for the economic accounting in Europe and North America. Today, leisure services account for 10% of the economy, but economic statisticians are only now extending their data. Interestingly, Thailand, a latecomer to economic accounting, has tourism data which are far superior to any European or American country. Economic models and scenarios have the same problem. Sectors which are small at the start of a scenario are not properly represented at the beginning, and, therefore, not throughout the entire scenario.

Technological detail suffers from a similar problem. Characterising current technology is one thing; characterising future technology is a different ballgame. An historical example may help. In the 19th century, people worried about the accumulation of horse manure, but that problem disappeared with the introduction of the internal combustion engine (Grübler, 1990). (We got other problems in return.) A model rich in technical detail would have characterised all the different types of transport by horse and carriage. A similar amount of detail on horseless transport was impossible to foresee, and perhaps the concept of horseless ground transport at scale was unforeseeable.

What could have been foreseen, however, is a continuation of technological progress in transport, always decreasing costs while increasing speed and capacity. As this trend has been there for as long as we know, there was little reason then (and there is little reason now) to assume it would break. Similarly, people's demand for travel seems to be insatiable, and is limited only by time and cost (Grübler, 1990).

If need be, economic models and scenarios can also distinguish between age and income classes (Dalton et al., forthcoming, van Heerden et al., 2006). Indeed, this is standard for tax analysis. However, every new dimension will require that a new dynamic be added to the model – often implying that the results become more speculative. For long run scenarios, a cruder resolution may be preferable.

4 How to Build a Scenario?

As already indicated above, building scenarios combines art and science. Scenarios are about the future, for which we do not have data. Long-term scenarios cannot be falsified, and they will not be proven wrong before considerable time has passed. Some people have great difficulty with this, as it violates their identity as "scientists". Other people take this as license that anything is possible. Such people should find another occupation. In my opinion, scenarios should be non-implausible extrapolations of processes that are reasonably understood. In that sense, scenarios are subject to peer-review and are even reproducible.

When building scenarios, one should consider what the scenario is about; and what it will be used for. This may appear to be a trite remark, but too often people build scenarios about everything,[7] and for everything. However, if scenarios are to be helpful in decision analysis, they should focus on the variables which determine the success or failure of the decisions under consideration. Economic scenarios revolve around three questions:

- How rich will we be, and what will we do with that richness?
- What will we use and what will we emit?
- How vulnerable will we be?

The answer to these questions determine the extent and nature of global change, the extent and nature of the threats and opportunities contained in global change, the ease with and way in which resource use and emissions can be changed, as well as the goals for policy.

5 Economic Growth

Economic growth is driven by three processes: population, capital, and technology.

The effect of the size of the population on economic growth is simple: More people will produce and consume more. Both production and consumption are important. Having less people does not imply that all will be richer, or even on average. Having less people also implies that less would be made. The pie has to be shared with less people, but it is also smaller (Birdsall et al., 2001).

The composition of the population is also important. The extraordinary economic growth in East Asia is partly explained by a labour force which is exceptionally large relative to the population. The aging of the people of Europe, China, and Japan may well slow down per capita economic growth (Birdsall et al., 2001).

[7] When building a scenario about everything, one necessarily encounters areas in which one's understanding is limited. This should be avoided. Instead, one should invite field experts, at least for advice. I have yet to find an area about which at least some people have not thought longer, harder and deeper than I have.

Capital accumulation is probably the best understood driver of economic growth (Solow, 1987). Capital is depreciated as age degrades its function, both economically and technically. Part of production is saved and invested in new capital. As workers have more capital at their disposal, they become more productive and command a higher salary.[8]

Technological progress is probably the most important driver of economic growth, accounting for perhaps 2/3 of observed growth; but is the also the least understood driver (Barro and Sala-i-Martin, 1995; Romer, 1996). Note that I used a broad definition of "technology", including both hardware and software. Economists prefer to define technological progress in terms of increases in productivity. Productivity would increase with education, experience, diffusion, innovation, and invention. At the same time, some knowledge would become obsolete; for example, the knowledge how to train and mind horses was once a crucial element of the transport system. Productivity improves partly by investment, particularly in education (training people) and innovation (research and development aimed at bringing existing products to the market). Here, inputs are relatively easy to measure, but outputs are not. It is easy to find out how much was spent on a college education, but assessing the value of that education is much harder. Similarly, expenditures on research and development can be measured (although data are suspect as R&D is often tax-deductible so that companies would tend to overstate spending), but the return on R&D depends on the counterfactual "how would the company have fared without the R&D?" Experience (doing things in a smarter way through repetition) and diffusion (copying from smarter peers) is even more intangible, also because average productivity increases as the least productive companies go out of business. Invention (new products and methods) is the most difficult. Putting a larger number of engineers to work on a problem would mean that existing ideas get better faster. New ideas, however, require a flash of genius.

The three fundamental drivers of economic growth are not independent. Less developed countries can grow very fast by accumulating capital, educating their work force, and copying methods and products from more advanced economies. As there are limits on how much existing products can improve, and how much better people can be educated, invention is most important in the most advanced economies (Funke and Strulik, 2000). Invention, and rapid advances in experience, are more likely with younger people (Romer, 1996).

Besides the question of how rich people in the future will be, there is the question about what will be done with this richness. It is relatively easy to imagine how people in poor countries will behave once they have reached middle incomes, and how people in middle-income countries will behave once they are rich. Obviously, they will not behave exactly the same as people in countries with mid or high incomes at present. They will not do so because their culture is different and because technologies and prices will be different.

---

[8] On average. Individual workers may be replaced by machinery.

Mobile phones, for instance, reduce the cost of communication for the currently poor in a way that the poor in the past never knew. Nonetheless, as people grow from poor to middle income to rich, they will spend less and less money on food and manufactured products, and more and more money on education, health care, and entertainment. This is not to say that there are no uncertainties – will China be like Japan or like the USA? –, however, the uncertainty is constrained by a mix of interpolation and extrapolation. For the currently rich, there is only extrapolation. Average incomes in Europe and North America are projected to exceed \$ 100,000 per person per year by 2100 (Nakicenovic and Swart, 2001). On current trends, most of this money would be spend on health care and holidays – which is implausible, but it is not known where the demand for these items satiates, and what money would be spend on instead.

6 Emissions

Trends in emissions can be understood with the Kaya Identity.[9] Suppose we are interested in the emissions of carbon dioxide, then the following holds

$$C \equiv \frac{CEY}{EYP}P \tag{1}$$

where C are emissions, E is energy use, Y is income, and P is number of people. Equation 1 implies that

$$\frac{\partial C}{\partial t} \equiv \frac{\partial \frac{C}{E}}{\partial t} + \frac{\partial \frac{E}{Y}}{\partial t} + \frac{\partial \frac{Y}{P}}{\partial t} + \frac{\partial P}{\partial t} \tag{2}$$

where t denotes time.

Population and income are discussed above. Energy intensity (energy use per value added) behaves fairly regularly: It gradually declines with technological progress and changes in the structure of the economy[10] (Lindmark, 2004; Tol et al., 2006). See Fig. 1 and 2. These are not just empirical regularities. Energy is a cost, and people and companies will economise on their energy use. As people grow richer, their consumption dematerialises – again reducing energy use.

Carbon intensity is a different story. It depends on the composition of the energy sector, that is, which sources of energy are used for what. This in turn depends on resources, prices and policies (Nakicenovic and Swart, 2001). For instance, fossil oil is the dominant energy source for transport. At some point in this century, conventional sources of oil will be exhausted. Will it be replaced by unconventional oil (tar sands, shale oil, heavy oil), by liquefied

[9] To my knowledge, Yoichi Kaya of RITE, Kyoto, never published a paper on this.
[10] The structure of the economy is the composition of production and consumption, that is, the share in agriculture, manufacturing, and services.

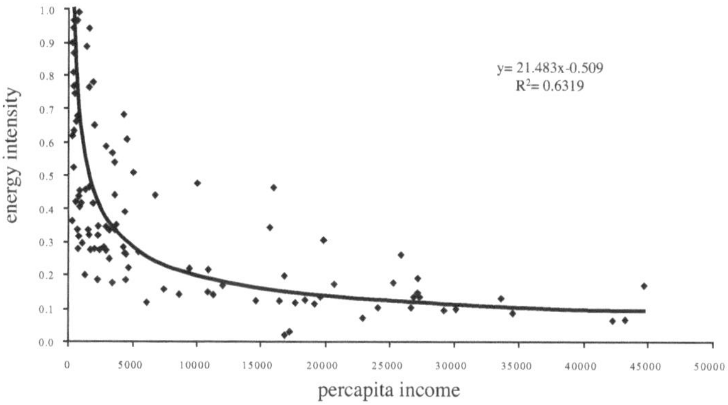

Fig. 1. Energy intensity (energy use per value added) and per capita income across the world. Source: WRI (2005)

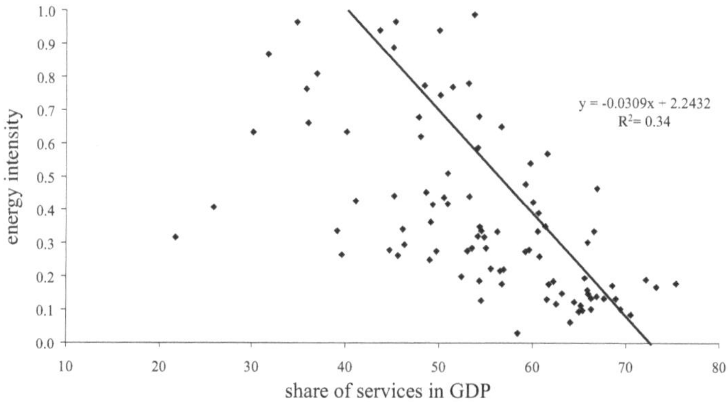

Fig. 2. Energy intensity (energy use per value added) and the share of services in GDP (an indicator of the structure of the economy) across the world. Source: WRI (2005)

coal, by biofuels, or by hydrogen? As this is a trend break, past experience is no guide. The only thing that is known is that the trend will break, perhaps later as resources run out or perhaps earlier as the growing market power of the Middle East is seen as threat. The economics of some alternatives may seem attractive at present – e.g., tar sands and biofuels – but this may change if the other aspects are taken into consideration – e.g., the problems of open pit mining of tar sands, and competition between energy crops and food crops. Here, projections are truly scenarios.

7 Vulnerability

Vulnerability (to climate change and other phenomena) obviously differs between countries, partly because of differences in development. Therefore, one should expect future vulnerability to be different from todays. Such a broad-brush statement is not very helpful, however. Vulnerability is not a simple concept either. People are not vulnerable, but vulnerable to something for a reason (Tol and Yohe, 2007). Only if one understands the reason, can one make projections of how specific vulnerabilities can change.

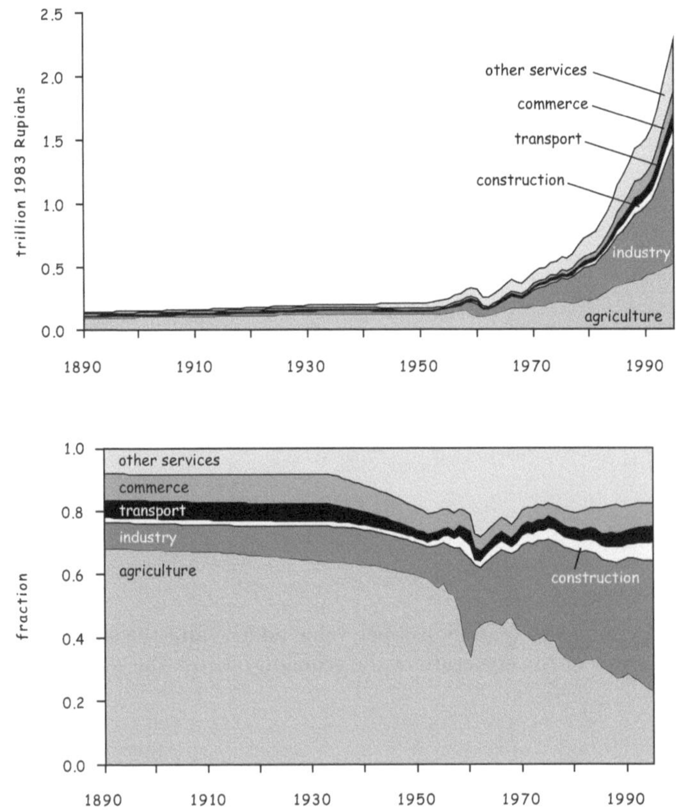

Fig. 3. The structure of the Chinese economy, in absolute (*top*) and relative (*bottom*) terms. Source: Maddison (1998)

Economic growth is a major driver of development. Economic growth reduces some vulnerabilities. For instance, the share of agriculture in the total economy shrinks as people grow richer. See Fig. 3 for China; Figure 4 for Indonesia. Economic impacts on agriculture will therefore shrink as well. In-

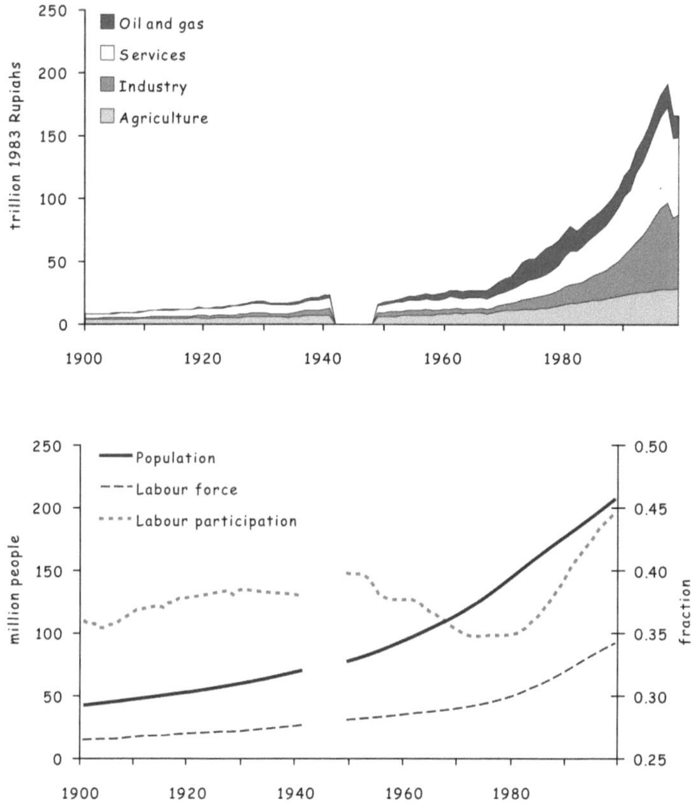

Fig. 4. The structure of the Indonesian economy (*top panel*) and the labour participation rate of the Indonesian population (*bottom panel*). Source: van der Eng (2002)

fectious diseases such as malaria and diarrhoea particularly afflict the poor, and would gradually disappear. On the other hand, cardiovascular diseases tend to be more prevalent in richer populations, because these people are older and fatter (Tol, 2002).

Some of these relationships are reasonably regular, and can therefore be modelled (Yohe and Tol, 2002). Figure 5 shows the correlation between per capita income on the one hand, and the probability of being affected by a weather-related natural disaster on the other hand. Richer countries are less vulnerable. But average weather is not the only thing that matters. Figure 5 also shows the relationship between the Gini coefficient (a measure of income distribution) and natural disasters. More egalitarian countries are less vulnerable. In the same analysis, it is shown that Buddhist countries are more vulnerable, and Muslim countries less (also when controlled for other

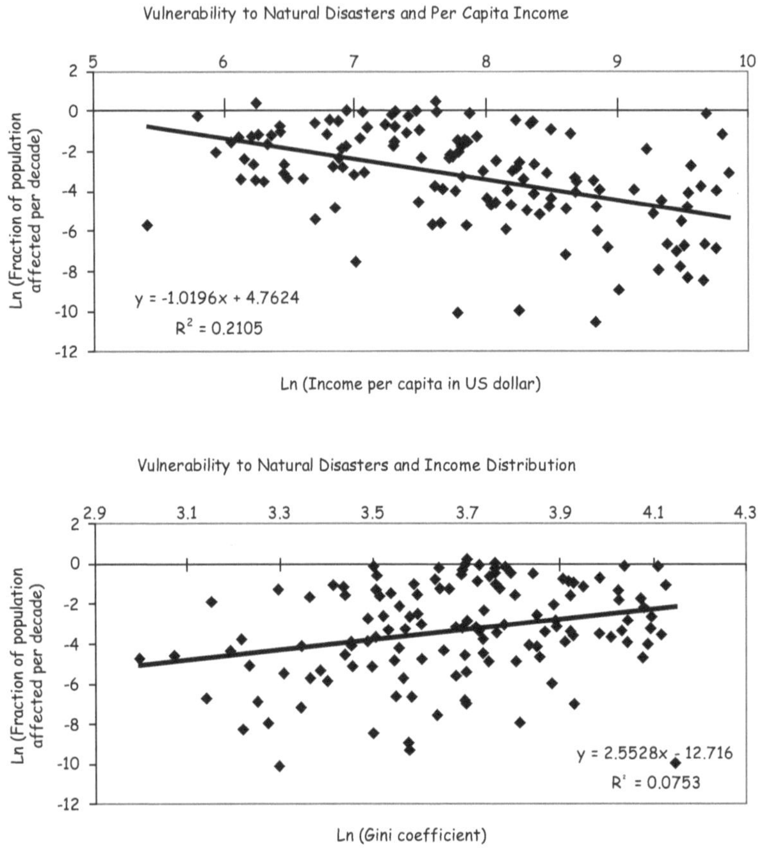

Fig. 5. The probability of being affected by a natural disaster versus per capita income (*top panel*) and income distribution (*bottom panel*) across the world in 1995. Source: Yohe and Tol (2002)

variables); democracy and civil liberties imply higher vulnerability, perhaps because safety measures require the strict, heavy-handed implementation that is so difficult in participatory democracies.

Some relationships are more irregular. Take, for instance, the statutory floods risks along the North Sea. Although per capita income is roughly equal and natural conditions very similar, protection standards vary from 1/100 in Flanders to 1/10,000 in Holland. Also in Germany, standards vary considerable between the states of Lower Saxony and Schleswig–Holstein. These differences can be understood only with an in-depth knowledge of the politics of flood protection.

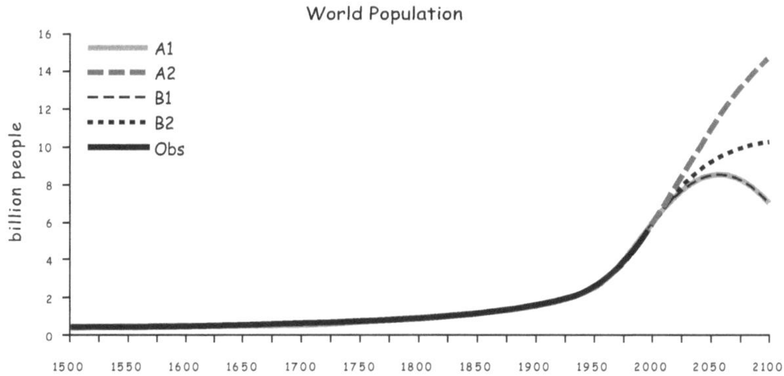

Fig. 6. World population. Sources: Maddison (2001) and Nakicenovic and Swart (2001)

8 Empirical Validity

Despite the difficulties, there are certain empirical regularities that scenarios should take into account. Figure 6 shows the world population as observed and as projected by the SRES scenarios of the IPCC. In the short term, the projections follow past trends. In the longer term, each scenario breaks the trend. This is plausible. Rapid population growth is an anomaly, brought about by advances in medicine and hygiene. High fertility is not a fundamental desire, but rather a response to high infant mortality. It is hard to imagine that developing countries will be rich but maintain high fertility; in fact, the two are contradictory, as one cannot be rich without proper education, and educating children is expensive (Galor and Weil, 1996, 1999).

Figure 7 shows observations and projections of per capita income for the USA, China, and Sub-Saharan Africa. For the USA, the SRES projections are in line with past experience. For China, the projections are in line with recent experience, but anomalous when compared to the longer term. This is fine for a single scenario, but all SRES scenarios share this characteristic. However, one can easily imagine a scenario in which China's aging population draws growth down; or a scenario in which democratisation poses bounds on China's ruthless capitalism; or even a scenario in which China descends into civil war.

For Sub-Saharan Africa, the SRES projections are clearly at odds with historical observations. The sparse data suggest that average per capita income was not much higher in 1960 than it was in 1500, and has fallen since; but all SRES scenarios have rapid economic growth for the future. Although one would hope that Africa will finally start to grow; and while comparisons to other continents suggests that it will eventually; there is little reason to assume that Africa will start to grow soon; indeed, the economic growth data

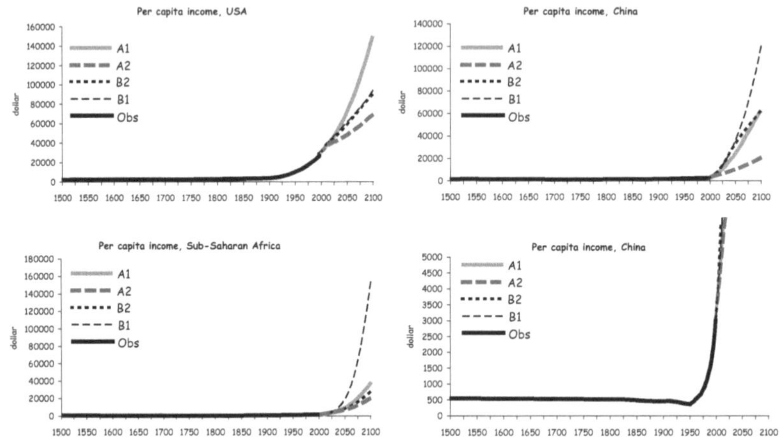

Fig. 7. Per capita income in the USA, Sub-Saharan Africa and China (shown at two different scales). Sources: Maddison (2001) and Nakicenovic and Swart (2001)

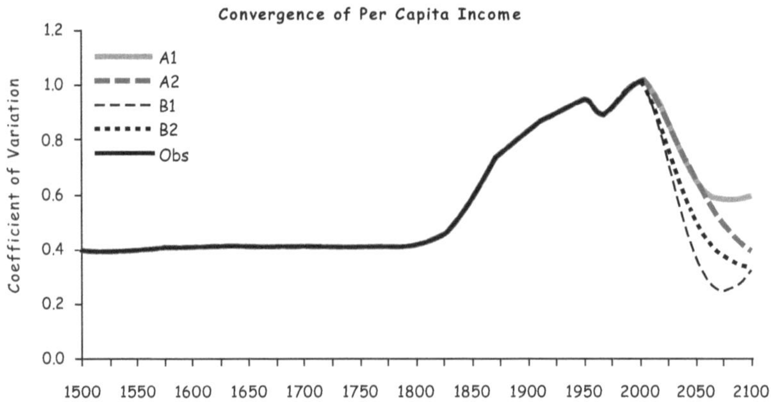

Fig. 8. The coefficient of variation of per capita income in the world. Sources: Maddison (2001) and Nakicenovic and Swart (2001)

which have become available since the publication of SRES discredit these assumptions. As always, some parts of Africa are growing economically, while other parts are in reverse.

Figure 8 confirms this. Since the start of the industrial revolution, the income distribution has relentlessly tended towards more inequality. This trend is reversed in each SRES scenario; not eventually, but right at the start of the projection.

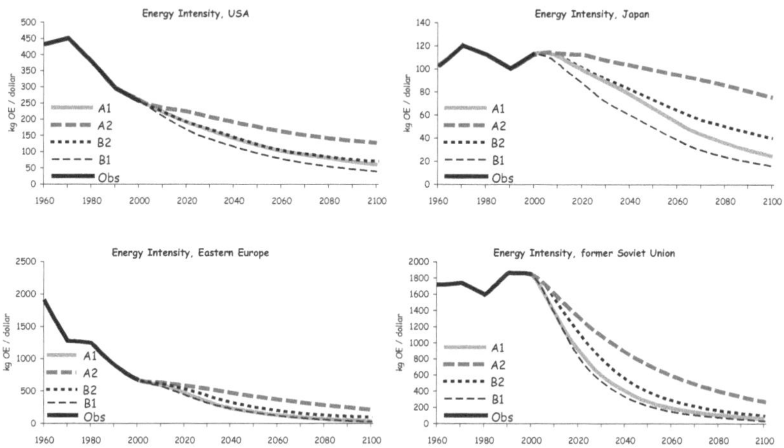

Fig. 9. The energy intensity of income in the USA, Japan, Eastern Europe and the former Soviet Union. Sources: IEA (2005), WRI (2005) and Nakicenovic and Swart (2001)

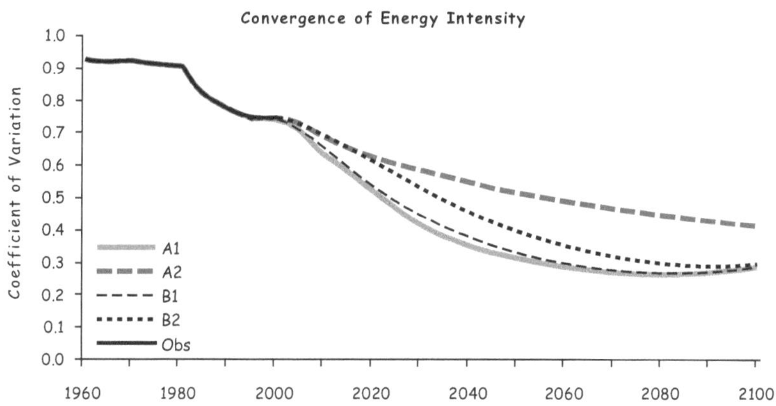

Fig. 10. The coefficient of variation of energy intensity in the world. Sources: IEA (2005), WRI (2005) and Nakicenovic and Swart (2001)

Figure 9 shows energy intensities, that is, the energy used per income generated. Unfortunately, long term data are hard to come by. Comparing the data from 1960 to 2000 to the scenarios, one cannot say the projections are implausible. It is clear, however, that the range of past experience is wider

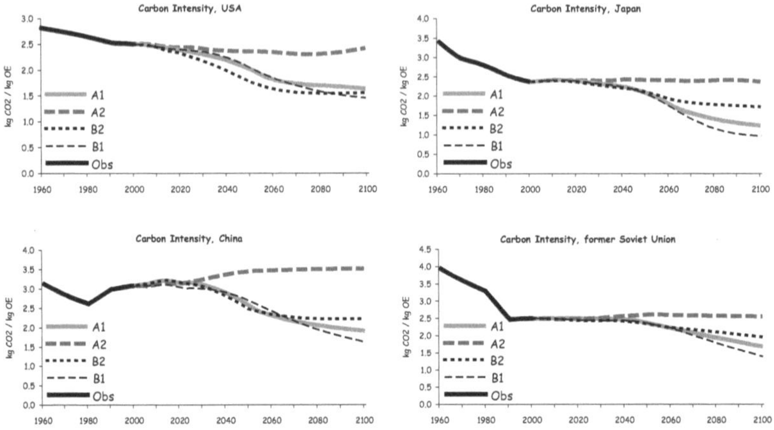

Fig. 11. The emission intensity of energy use in the USA, Japan, Eastern Europe and the former Soviet Union. Sources: IEA (2005), WRI (2005) and Nakicenovic and Swart (2001)

than the range of future projections. The same holds for the convergence of energy intensity, shown in Fig. 10, and for carbon intensity, shown in Fig. 11. The convergence of carbon intensity, shown in Fig. 12, shows an interesting pattern. Carbon intensities have converged in the last four decades, and this process is projected to continue. However, this trend is reversed and may well happen. Also, the trend is reversed in the four scenarios around the same time (2020); the scenarios can hardly be called "alternative" scenarios in this respect, as they behave almost exactly the same.

The above exercise shows, that although scenarios cannot be falsified or validated, they can be tested against data nonetheless. The past is an imperfect guide to the future, but the range of past experiences is wide enough to calibrate what is implausible and what is not. One would ignore that experience at one's peril.

9 Conclusion

Scenarios are forecasts conditioned on difficult-to-predict boundary conditions. In a mathematical sense, scenarios are conditional forecasts – but in a semantic sense, scenarios are different because the forecast or the conditions have methodologically weak spots. The long-term research agenda should be to turn scenarios into proper conditional forecasts – the only free variable should be the decision the analysis seeks to inform. At present, we are a long way removed from this. For now, scenarios are not-implausible descriptions of alternative futures. Scenarios are best used to test the robustness of policies. Economic scenarios are scenarios of economic variables.

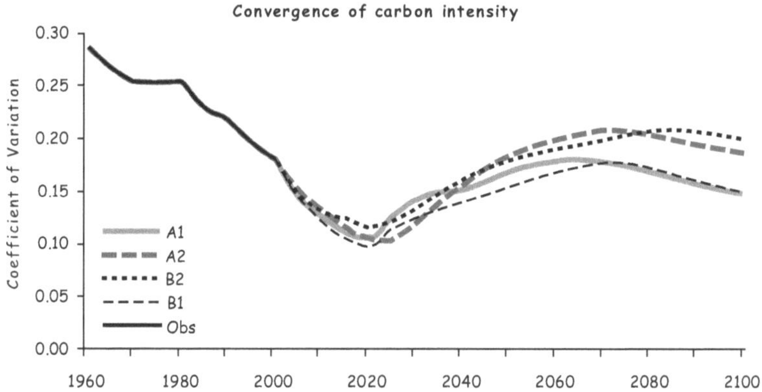

Fig. 12. The coefficient of variation of energy intensity in the world. Sources: IEA (2005), WRI (2005) and Nakicenovic and Swart (2001)

In scenario development, there is a trade-off between adding process knowledge and adding detail. The former increases the quality of a scenario, while the latter increases the uncertainty. In the very long term, broad trends on large scales are easier to foresee than details. Although scenarios cannot be validated, their implausibility can be tested by plotting them against observations.

With the current state of knowledge, we cannot conditionally predict future economic growth, future emissions, or future vulnerability to climate change. We can, however, use empirical regularities and theoretical understanding to bound future growth rates, emissions, and vulnerabilities.

Acknowledgements

Hans von Storch had constructive comments which helped to improve the paper. Financial support by the Hamburg University Innovation Fund, the Princeton Environmental Institute, and the ESRI Energy Policy Research Centre is gratefully acknowledged.

References

Barrow RJ, Sala-i-Martin X (1995) Economic Growth, McGrawHill, New York

Birdsall N, Kelley AC, Sinding SW (eds) (2001) Population Matters – Demographic Change, Economic Growth, and Poverty in the Developing World Oxford. University Press, Oxford

Boehringer C, Löschel A (eds) (2003) Empirical Modeling of the Economy and the Environment. Physica-Verlag, Heidelberg New York

Brakman S, Garretsen H, van Marrewijk C (2001) An Introduction to Geographical Economics – Trade, Location and Growth. Cambridge University Press, Cambridge UK

Dalton MG, O'Neill B, Prskawetz A, Jiang L, Pitkin J (forthcoming) Population Aging and Future Carbon Emissions in the United States. Energy Economics

Ericksen NJ (1975) Scenario Methodology in Natural Hazards Research. Institute of Behavioral Science, University of Colorado, Boulder, NSF-RA-E-75-010

Fujita M, Krugman P, Venables AJ (1999) The Spatial Economy – Cities, Regions, and International Trade. MIT Press, Cambridge London

Funke M, Strulik H (2000) On endogenous growth with physical capital, human capital and product variety. Europ Econ Rev 44:491–515

Galor O, Weil DN (1999) From Malthusian Stagnation to Modern Growth. Am Econ Rev 89,2:150–154

Galor O, Weil DN (1996) The Gender Gap, Fertility, and Growth. Am Econ Rev 86,3:374–387

Glimcher PW (2003) Decisions, Uncertainty, and the Brain – The Science of Neuroeconomics. MIT Press, Cambridge London

Grübler A (1990) Rise and Fall of Infrastructures: Dynamics of Evolution and Technological Change in Transport. Physica Verlag, Heidelberg

IEA (2005) World Energy Statistics and Balances, International Energy Agency, http://hermia.sourceoecd.com/vl=6413579/cl=13/nw=1/rpsv/statistic/s35_about.htm?jnlissn=16834240

IPCC (2001) Climate Change 2001: The Scientific Basis – Contribution of Working Group 1 to the Third Assessment Report of the Intergovernmental Panel on Climate Change. Cambrige University Press, Cambridge

Lempert RJ, Nakicenovic N, Sarewitz D, Schlesinger ME (2004) Characterizing climate-change uncertainties for decision-makers. Clim Change 65:1–9

Lindmark M (2004) Patterns of Historical CO_2 Intensity Transitions among High and Low Income Countries. Explor in Econ Hist 41:426–447

Maddison A (1998) Chinese Economic Performance in the Long Run. OECD, Paris

Maddison A (2001) The World Economy: A Millennial Perspective. OECD Publications, Paris

Nakicenovic N Swart RJ (eds) (2001) IPCC Special Report on Emissions Scenarios Cambrigde University Press, Cambridge

Nakicenovic N, Victor N, Morita T (1998) Emissions Scenarios Database and Review of Scenarios. Mitig and Adapt Strategies for Glob Change 3,2–4:95–131

Postma Th JBM, Liebl F (2005) How to improve scenario analysis as a strategic management tool? Technol Forecasting and Soc Change 72:161–173

Romer D (1996) Advanced Macroeconomics. McGraw-Hill, New York

Samuelson PA, Nordhaus WD, Sutton L (ed) (2001) Macroeconomics. 17th ed, McGraw-Hill

Solow RM (1987) Growth Theory – An Exposition (The Radcliffe Lectures Delivered in the University of Warwick 1969, Nobel Prize Lecture 1987). Oxford University Press, New York

Strzepek KM, Yates DN, Yohe GW, Tol RSJ, Mader N (2001) Constructing "not implausible" climate and economic scenarios for Egypt. Integr Assessm 2:139–157

Tol RSJ (2002) New Estimates of the Damage Costs of Climate Change, Part II: Dynamic Estimates. Env and Resource Econ 21,1:135–160

Tol RSJ, Pacala SW, Socolow RH (2006) Understanding long-term energy use and carbon emissions in the USA. Research unit Sustainability and Global Change FNU-100, Hamburg University and Centre for Marine and Atmospheric Science, Hamburg

Tol RSJ, Yohe GW (2007) The Weakest Link Hypothesis for Adaptive Capacity: An Empirical Test. Glob Env Change 17:218–227

Tschirhart J (2000) General Equilibrium of an Ecosystem. J Theor Biol 203:13–32

van der Eng P (2002) Indonesia's Growth Performance in the Twentieth Century. In: Maddison A, Prasada Rao DS, Shepherd WF (2002) The Asian Economies in the Twentieth Century, Edward Elgar, Cheltenham

van Heerden J, Gerlagh R, Blignaut JN, Horridge M, Hess S, Mabugu R, Mabugu M (2006) Searching for Triple Dividends in South Africa: Fighting CO_2 Pollution and Poverty while Promoting Growth. Energy J 27,2:113–141

van Notten PhWF, Sleegers AM, van Asselt MBA (2005) The future shocks: On discontinuity and scenario development. Technol Forecasting and Soc Change 72:175–194

Weibull JW (1995) Evolutionary Game Theory MIT Press, Cambridge

WRI (2005) EarthTrends Environmental Information. http://earthtrends.wri.org/

Yohe GW (1991) Selecting Interesting Scenarios with which to Analyze Policy Response to Potential Climate Change. Clim Res 1:169–177

Yohe GW, Tol RSJ (2002) Indicators for Social and Economic Coping Capacity – Moving Towards a Working Definition of Adaptive Capacity. Glob Env Change:12,1:25–40

Yohe GW, Jacobsen M, Gapotchenko T (1999) Spanning "Not-Implausible" Futures to Assess Relative Vulnerability to Climate Change and Climate Variability. Glob Env Change 9,3:233–250

Climate Change: Regulating the Unknown

Richard S.J. Tol[1,2,3]

[1] Economic and Social Research Institute, Dublin, Ireland
[2] Institute for Environmental Studies, Vrije Universiteit, Amsterdam, The Netherlands
[3] Engineering and Public Policy, Carnegie Mellon University, Pittsburgh, PA, USA

1 Introduction

Everything about climate change is uncertain: Future emissions, past emissions, current emissions; the carbon cycle; the climate; impacts of climate change; costs of emission abatement; and the effect of abatement policies. This is no reason for despair. Uncertainty is a central feature in many policy domains. A suite of methods for policy analysis and advice under uncertain conditions have been developed, and are being applied daily, for example in the financial sector. The literature on uncertainty is extensive (Chichilnisky, 2000; Dixit and Pindyck, 1994; Morgan and Henrion, 1990; Pratt et al., 1995), and so is its application to climate change (Allen et al., 2000; Baker, 2005; CBO, 2005; Eismont and Welsch, 1996; Forest et al., 2002; Hammitt, 1995; Harvey, 1996a,b; Keller et al., 2004; Kolstad, 1996; Lempert et al., 2004; Manne and Richels, 1992; Pate-Cornell, 1996; Pizer, 1999; Tol, 2003; Ulph and Maddison, 1997; Webster et al., 2003; Yohe, 1997; Zapert et al., 1998)

Yet, in climate policy, two responses to uncertainty dominate. Some deny that there is a climate problem. Others deny that there is uncertainty about the problem.

Denying that there is a problem is just dumb. It is in contrast with the precautionary principle (Dovers and Handmer, 1995; Martin, 1997), both in its Anglo-Saxon version (uncertainty is no excuse for inaction) and in its German version (better safe than sorry). A reasonable decision takes into account the downside risks of that decision and its alternatives. One has to read the evidence about climate change and its impacts very selectively to conclude, with any degree of confidence, that there is no problem. It is hard to suppress the suspicion that people want to believe this, and this against all reason.

Denying that there is uncertainty is not particularly smart either. Decisions which are based on partial evidence may be fragile. The option to revise policy in the light of new knowledge is valuable, particularly if uncertainties are large, time-scales long and learning substantial (Manne and Richels, 1992). Yet, in climate change, policy makers seem to be in a rush to settle things for which the scientific basis is still lacking. All aspects of climate change are uncertain, but some aspects are so uncertain that it is better to wait for additional information. Indeed, decisions on climate policy are often made without any policy analysis, also in cases where a few months of reflection would have prevented policy mistakes.

Aviation emissions are a recent example. I wrote a paper on carbon taxes on aircraft emissions in September 2006 (Tol, 2007), and felt obliged to portray this as a remote prospect – there had been extensive discussion, but a policy consensus seemed remote. In December 2006, the decision was made to include aviation emissions in the European trading system for carbon dioxide. At that time, there were only two reports which advised the European Commission on this matter (Wit et al., 2002, 2005), and only three other papers (Michaelis, 1997; Olsthoorn, 2002; Tuinstra et al., 2005). The policy to be implemented differs substantially from earlier policy proposals; and the accompanying assessment that accompanies (CEC, 2006) does not include any concrete analysis. A post-hoc analysis by Tol and FitzGerald (2007) reveals that the policy is ineffective and expensive and shows that a few minor changes would have lead to much larger emission reductions for the same cost. The European Council could have announced an intention, invited comments, and made a better informed decision in December 2007. Instead, it rushed to regulation in December 2006.

Climate change is of course a terrible problem for politicians; for reelection, tangible results must be had within a year or two. To solve the climate problem, policy would need to be steady for decades. As a consequence, decisions are made without much reason. This would be fine if revisions were possible, but this is often hard in international diplomacy.

In this paper, I review five other examples in which decisions were too hasty. In the first examples, this may be because of clumsy diplomacy. The latter examples reveal incompetence or bad intentions. The five examples are global warming potentials, multi-gas emission reduction, ancillary benefits, the social costs of carbon, and the European long-term climate policy target.

2 Global Warming Potentials

The global warming potentials (GWPs) allow one to express emissions of each greenhouse gas in a carbon dioxide equivalent. The IPCC (Ramaswamy et al., 2001) defined the GWP of, say, methane as the ratio of the time integral from 2000 to 2100 of the incremental radiative forcing due to a slight increase in methane emission to the same integral for carbon dioxide emission. The GWP is a purely physical concept. Some people find this an attractive property, and it would be if climate change were a physical problem. Even so, the time horizon (100 years) is arbitrary; and as radiative forcing is non-linear in concentration, the GWPs are sensitive to the baseline scenario (Smith and

Wigley, 2000a,b; Shine et al., 2005) and specifically, the GWP assumes that greenhouse gas concentrations will remain constant, the least likely scenario. These properties undermine the "objective" nature of the GWP. However, the time horizon is fixed by convention, while not many are worried about the arbitrary choice of scenarios.

The main application of the GWP, however, is in policy: The GWP establishes the exchange rate between greenhouse gases. Policies should have a purpose. If the purpose of policy is re-election or promotion, and the probability of re-election or promotion is increased twice as much by methane abatement than by carbon dioxide abatement, then the appropriate GWP for methane is two. Of course, politicians state loftier goals for policy. The stated aim of international climate policy is to stabilize atmospheric concentrations of greenhouse gases, presumably in some distant future. Methane is a short-lived gas, and current emission reduction therefore does not contribute to long-term stabilization. The GWP of methane should therefore be low, if not zero. If one is interested in atmospheric stabilization at the lowest possible costs, then the appropriate exchange rate is the ratio of the shadow prices (Godal and Fuglestvedt, 2002; Tol et al., 2003). Manne and Richels (2001) computed this and found that the GWP of methane should start much lower than the 21 adopted in the Kyoto Protocol and end much higher.

If one thinks that the aim of climate policy is to avoid impacts of climate change, then the appropriate exchange rate is the ratio of impacts caused by the respective emissions, or rather the ratio of marginal impacts (Fankhauser, 1995; Kandlikar, 1995, 1996; Reilly and Richards, 1993; Schmalensee, 1993; Sygna et al., 2002; Tol, 1999). In this case, methane is more important than that suggested by the GWP. There are three reasons for this. First, because of time discounting, the short-term impacts of methane are more important than the long-term impacts of carbon dioxide. Second, impacts are partly driven by the rate of warming, and the next few decades will witness faster warming than the more remote future. Methane accelerates the near-term rate of warming more than carbon dioxide. Third, carbon dioxide has benefits in the form of CO_2 fertilization that methane does not have.

So, there are arguments to believe that the GWP of methane is too low and too high. One might argue that the GWP is a compromise. This is not true [4]. A shower may alternate between too hot and too cold; although the average water temperature is right, the experience is unpleasant nonetheless. More importantly, politics is about making choices and accepting the consequences.

[4] It is not true in logical sense either. Working Group 1 of the IPCC has ignored all non-physical literature on the subject, arguing that this is Working Group 3 territory, which is apparently beyond the comprehension of physicists. At the same time, Working Group 3, which harbours the appropriate expertise, refuses to get involved, perhaps because fixing the GWP would have real political and economic implications. The relationship between WG1 and WG3 is problematic at a personal level too.

Using the average between two alternatives is avoiding rather than confronting the issue.

The dynamics of the GWP issue are confused. Politicians hide behind the IPCC, saying that this is a scientific issue. Indeed, the Kyoto Protocol explicitly delegates the issue to the IPCC. IPCC officials argue that the GWPs should not be changed lightly, as they have been adopted by policy makers. This circular reasoning freezes the unfortunate status quo. Either policy makers should charge the IPCC to reconsider the issue, or the IPCC should assume their academic duty to give unsolicited advice.

3 Multi-Gas Emission Reduction

In the Kyoto Protocol, six "gases" are regulated: carbon dioxide (CO_2), methane (CH_4), nitrous oxide (N_2O), sulphur hexafluoride (SF_6), halofluorocarbons (HFCs), and perfluorocarbons (PFCs) (the last two are classes of gases.) In principle, this is sound. The six gases constitute virtually all of the anthropogenic greenhouse gases, while most of the remaining gases (chlorofluorocarbons and bromides) will be phased out under the Montreal Protocol for the protection of the ozone layer. More options for emission reduction imply greater flexibility, lower costs, and less resistance to climate policy. That is true in theory, but reality may be different.

At the time the Kyoto Protocol was negotiated, the literature on emission reduction for non-CO_2 greenhouse gases was very thin (Tol et al., 2003). This has now changed (Van Vuuren et al., 2006), but there is hardly any knowledge on policy instruments[5]. Negotiators thus agreed on doing something, not knowing how, or what it would cost. Policy makers rushed ahead of the science. This is peculiar for a democracy, in which politicians are supposedly accountable to the people, and therefore should at least be able to explain why certain choices were made. Without any research done, such explanations cannot be given.

Combined with the high GWP for methane (see above), this has led to some peculiar phenomena. Total greenhouse gas emissions are sensitive to the animal stock. In the early 1990s, New Zealand mutton exports dived, taking methane emissions with them [6]. The same may happen in Ireland, as the reform of the Common Agricultural Policy is likely to replace sheep farming. Agriculture in developing countries is also an attractive option for emission reduction, through the Clean Development Mechanism (CDM). The price of methane is high enough to disregard the implications for food production. The market for SF6 is most distorted; because of the high GWP, reducing emissions

[5] For example, John Reilly of MIT has floated the idea of a methane tax, to be levied on millions of small rice growers and cow farmers. This work has not appeared in print.

[6] Under the UNFCCC, exporters are responsible for emissions, rather than importers, with the exception of fossil fuels.

is far more lucrative than their industrial application. Worryingly, data on SF6 emissions are hard to obtain and are of poor quality, so that fictitious emission reduction is lucrative, with a low probability of being caught. HFCs are most astounding. HFCs are produced only to bring in CDM money to avoid emissions (Harvey, 2007), a practice that takes away resources from real emission reduction (Schwank, 2004).

In hindsight, these effects could have been predicted – if only politicians had granted themselves enough time to ask the right questions of their experts. But as with the GWPs and aviation emissions, this is an example of clumsiness borne from haste. I will now turn to more serious examples.

4 Ancillary Benefits

Developing countries have no obligations to reduce their greenhouse gas emissions. If climate change is to be substantially slowed down, this will have to change. The governments of most developing countries know this, but they insist on developed countries taking the lead – after all, developed countries are responsible for the lion's share of historical emissions. Developed countries would like to shift part of the burden to developing countries, or at least part of the implementation. One tactic is to stress the so-called ancillary benefits of greenhouse gas emission reduction. Carbon dioxide emissions originate from the burning of fossil fuel, which also causes indoor air pollution, urban air pollution, and acidification; fossil fuel use is also a reason for energy insecurity, politically and economically. Reducing carbon dioxide emissions would also alleviate these problems, and these are known as the ancillary or secondary benefits of greenhouse gas emission reduction (Ekins, 1996; Working Group on Public Health and Fossil Fuel Combustion, 1997; Holmes and Ellis, 1999; Hourcade et al., 2001; Burtraw et al., 2003, Dessus and O'Connor, 2003; van Vuuren et al., 2006).

Air pollution is a big problem, particularly in the rapidly growing economies of South and East Asia. Although climate policy would reduce air pollution, it is not true that climate policy is the best option for reducing air pollution. Other policies would reduce the emissions of air pollutants more effectively, and would be much cheaper. This is particularly true for filters on exhaust gases. Filters typically increase carbon dioxide emissions, as energy efficiency falls. In many developing countries, air pollution is seen as a bigger problem than climate change. Climate change is in fact an ancillary non-benefit of air pollution policy; and improved air quality is not an ancillary benefit of climate policy.

Similarly, energy security may be a problem – dependence on foreign energy sources may prevent engaging in or winning a war, while shocks in energy prices may cause economic recessions (Bernanke et al., 1997). However, besides improved energy efficiency, the best solution for energy security may be domestic coal (in China, India, and the USA), rather than wind or solar power. Nuclear power would reduce energy imports and greenhouse gas emissions, but the enhanced risk of proliferation has unclear implications for international security.

An economist would immediately recognize an attempt to solve air pollution or energy security by climate policy as a second-best solution, and would know that second-best solutions are sub-optimal and are to be avoided if at all possible. Although second-best solutions are often better than no solution at all, air pollution and energy security should be tackled head-on, rather than through the back-door of climate policy. Nonetheless, the OECD has hired high-powered economists to quantify the ancillary benefits of climate policy so as to convince developing countries to adopt greenhouse gas emission reduction targets (e.g., Dessus and O'Conner, 2003). Here, scientific knowledge is knowingly abused in an attempt to manipulate the policies of other countries. This happens all the time, but the manipulated countries have in fact only limited access to expert knowledge.

5 The Social Cost of Carbon

The social cost of carbon is the British term for the marginal damage costs of carbon dioxide emissions, or the Pigou tax. It is a measure of the seriousness of climate change impacts, and a measure of the desired intensity of abatement policy. It is how much climate policy should cost, how high the carbon tax should be, or how expensive carbon permits should be.

The social cost of carbon is the additional damage done by a small increase in carbon dioxide emissions. Estimates depend on the atmospheric life-time of carbon dioxide, and on the response of atmosphere and ocean to changes in radiative forcing. Estimates also depend on the impacts of climate change, and how they are measured and aggregated over countries and sectors. Scenarios of emissions, population, and economic activity are important. Last but not least, the incremental damages have to be aggregated over time.

In 2003, HM Treasury (Clarkson and Deyes, 2003) published an estimate of the social cost of carbon: £70/tC, rising by £1/tC a year. The UK Government is committed to climate policy and to the Kyoto Protocol. At the same time, it is committed to cost-benefit analysis. The study by HM Treasury was part of an attempt to do a cost-benefit analysis on British climate policy. With a carbon tax starting at £70/tC, the UK would meet its Kyoto targets; and also its self-imposed targets for later dates. It is always convenient when an analysis confirms that the right choice has been made.

The Clarkson and Deyes (2003) study was a literature survey. Pearce (2003) and Tol (2005a) survey the same literature, and conclude that the best guess social cost is an order of magnitude smaller [7]. In fact, the three studies cover much the same material – and reading Clarkson and Davis (2003) one would expect a similar conclusion. Indeed, Pearce (2003) emphasizes that he agrees with the main text of Clarkson and Davis (2003); that he strongly disagrees with their conclusion and that the text and conclusion seem to be disconnected.

An innocent explanation is that a zero was accidentally added to the summary of Clarkson and Davis (2003). Another explanation is that the summary was altered under political pressure.

Whatever the truth may be, Pearce's (2003) paper has caused HM Treasury to issue a new study (Downing et al., 2005; Watkiss et al., 2005) [8] – which has not gained prominence as the House of Lords (2005) report has led to yet another inquiry: the Stern Review (Stern et al., 2006). Political hyperactivity and continuous re-assessment of an only slowly expanding knowledge base are a fact of life in climate change.

If Clarkson and Deyes (2003) is a bit odd, the Stern Review is very strange. The conclusion of Clarkson and Davis (2003) missed the average in the surveyed literature by an order of magnitude. The Stern Review managed to be two orders of magnitude off. The Stern Review does not present new data, new methods, or new models; it is a survey. Yet, a selective reading of the literature, an extreme choice of parameters, and a compilation of basic errors led to conclusions which are in sharp contrast to other studies (Tol, 2006; Tol and Yohe, 2006; Yohe, 2006). The politics of the Stern Review are obscure. The sponsors, Tony Blair and Gordon Brown, rapidly distanced themselves from the Stern Review; and Nick Stern resigned.

The final estimate of the social cost of carbon depend on a large number of assumptions, and one can massage the number up or down if so required. This is partly because scientific uncertainties are large and open to diverse interpretation. However, the social cost of carbon also depends on ethical choices, such as how much one cares about risk, about the future, or about people in other lands. Although the estimates are very uncertain, it is not true that anything goes. Some estimates are clearly less credible than others. On matters of ethics, a wide range of positions can be defended – but in a democratic country, the government cannot move too far from the opinion of the average voter. Having watched the deliberations in the UK from a safe but not too great a distance, it is clear that certain policy makers have forced the hand of experts. Helm (2005) speaks of the "tortuous attempts to reconcile the political targets with CBA approaches" and warns against the attitude of "starting with the conclusion and working back to find a set of consistent

[7] Note that £1/tC/yr increase does not follow from the literature; in fact, there were few studies on this published before 2003. In a later study, Tol (2005b) reaches the opposite conclusion.

[8] These studies halve the estimate to £35/tC; this number is as fake as the initial estimate, but it saves faces.

assumptions", an attitude which, he thinks, "pervades much of the policy process".

6 Europe's Long-Term Climate Policy Target

The European Council "believes that global average temperatures should not exceed 2 °C above pre-industrial level" (CEU, 1996, 2004). Separately, the governments of Germany, the Netherlands, Sweden and the UK have made this their policy goal as well. This is a stringent target. If the target is to be met with high confidence, CO_2 concentrations may have to be kept below 400 ppm (e.g., Meinshausen, 2005). The current concentration is 380 ppm, rising by 3 ppm per year. It is unclear whether meeting the 2 °C target is technically feasible; politically, it is very tough; economically, it is very costly.

The 2 °C target was originally proposed by the WBGU (1995), the Scientific Advisory Council Global Environmental Change of the German government. Two arguments are offered. First, this target would safeguard creation. This is a peculiar statement for scientific advisors of a secular government. The WBGU (1995) argues that it has never been warmer than 16.1 °C (global average), and that therefore it should not get warmer than 16.5 °C – roughly 2 °C above pre-industrial. This is a clear example of the naturalist fallacy. China has never been a democracy, and therefore it should never be a democracy. The WBGU (1995) asserts that there would be "drastic ecological impacts", without reference or further specification. In 1995, research into the ecological impacts of climate change was in its infancy; therefore, there was no scientific basis for this claim. One is often forced to make decisions based on incomplete information; but then one should revise the decision as new information becomes available. The second argument of the WBGU is based on intolerable costs. The WBGU states, but not argues, that an annual damage of 5% of GDP is just bearable. Pearce et al. (1996) estimate the annual damage of climate change at 1–2% of GDP. The WBGU states that this estimate is biased, and raises the impact to 5%. The bias is due to the omission of extreme weather events. Dorland et al. (1999) estimate the annual impact of an increase in storminess in Europe at 0.05%. Floods would have to increase by a factor 140,000 in the USA to reach 3% of GDP.

Although the WBGU (1995) defense is dubious at best[9], the 2 °C target was adopted by other governments and the EU as well. The European Commission (CEC, 2005a,b) seeks to defend the target. Like the UK Government, the European Commission has obliged itself to a cost-benefit analysis on any major policy initiative. CEC (2005a,b) claims to do that for climate policy.

[9] The WBGU (2003) offers new support. However, most of this is drawn from a report written by the climate director of Greenpeace International. As stated before, new information should imply new targets. Insight into the impact of climate change had grown manifold between WBGU (1995) and WBGU (2005).

The document is most peculiar. First, the impact of climate change is confused with the impact that would be avoided by emission reduction. Second, the estimates of Pearce et al. (1996) are quoted, but no study since then - and impact estimates have fallen since 1996 (Smith et al., 2001; Tol, 2005a). Third, emission reduction costs are counted to 2030 only, even though most of the costs would occur after 2030. Fourth, cost-benefit analysis requires the equation of marginal costs and benefits, a step that is omitted from CEC (2005a,b) [10]. Fifth, the report fails to review the extensive literature on cost-benefit analysis for climate policy, including the papers published in such journals as Science, the Economic Journal and the American Economic Review. If CEC (2005b) would be submitted as a term paper in an economics course, it would fail [11].

Where the UK Government tried to put a spin on research results (see previous section), the European Commission delivered shoddy work, adding insult to injury.

In both cases, policy makers are hiding behind "sounds-as-science". Instead, policy makers should base their results on sound science. Policy maker should also make clear that they are responsible for their decisions, and be prepared to defend the value judgements that are part of any decision.

Researchers – whether academics, consultants or officials – should stop providing excuses to politicians. Bjorn Lomborg was quickly accused of academic dishonesty, and had to defend his book "The Skeptical Environmentalist" to a committee of the Danish Academy. Some of the players in the issues reviewed in this section and the previous could just as easily have been accused on academic dishonesty, and perhaps they should have been.

7 Conclusion

Accepting the need for emission reduction is not the same as closing one's eyes for all evidence and rushing into abatement policies that may or may not work. Many decisions in climate policy have been made on the basis on faulty or incomplete science, on the basis of faulty interpretations of evidence, and on the basis of knowingly misinterpreted research. This is nothing new. Such is life in politics. However, climate policy is not for the current election cycle only. Climate policy is for the long haul, for decades or more. You can fool some of the people all of the time, and all of the people some of the time, but you cannot fool all of the people all of the time. A policy based on wrong science, with interpretations which cannot possibly be construed as "honest mistakes", only provides ammunition to the opposition to climate

[10] An earlier report makes the same mistake. When I pointed this out to the author while discussing a draft version, the author (an official of the European Commission) replied that he knew well the method was wrong, but this was the only way to get the desired answer.

[11] The current author is a full professor of environmental economics.

policy; and may increase the distrust of the electorate in their politicians and civil servants. This sort of tactics may work in the short run, but may lead to a backlash which hinders progress in the long run.

That is not to say that climate policy should wait until all scientific issues are settled. On the contrary, uncertainty is a good reason to push climate policy. However, scientific uncertainty and controversy do require that decisions are revised when uncertainties are reduced or controversies are lifted. Climate policy makers have demonstrated a distinct lack of willingness to reconsider previous decisions, probably because most decisions are painfully wrought compromises.

If decisions are hard to reverse, they should not be rushed. Particularly in Europe, politicians apparently feel the need to be seen to make decisions on climate policy – and lack of public scrutiny implies that bad decisions are tolerated. In such an atmosphere, any decision is better than no decision. This suggests that the quality of decision making will not improve until the public debate becomes less hysterical.

Climate policy analysts are overstretched by the rapid succession of policy initiatives and regulations. Still, they should stretch harder to add more ex ante policy analyses to their portfolio, anticipating policy proposals and preventing bad decisions.

Acknowledgements

Susanne Adams' comments helped to improve the paper substantially. Funding by the Hamburg University Innovation Fund, the Princeton Environmental Institute, and the ESRI Energy Policy Research Institute is gratefully acknowledged.

References

Allen MR, Stott PA, Mitchell JFB, Schnur R, Delworth TL (2000) Quantifying the uncertainty in forecasts of anthropogenic climate change. Nature 407:617-620

Baker E (2005) Uncertainty and learning in a strategic environment: global climate change. Res Energy Econ 27:19–40

Bernanke BS, Gertler M, Watson MW (1997) Systematic Monetary Policy and the Effects of Oil Price Shocks. Brookings Papers on Economic Activity 1:97-157

Burtraw D, Krupnick A, Palmer K, Paul A, Toman M, Bloyd C (2003) Ancillary benefits of reduced air pollution in the US from moderate greenhouse gas mitigation policies in the electricity sector. J Env Econ Manag 45:650–673

CBO (2005) Uncertainty in Analyzing Climate Change: Policy Implications, Congress of the United States, Congressional Budget Office, Washington, DC/USA.

CEC (2005a) Winning the Battle against Global Climate Change, Communication from the Commission to the Council, the European Parliament, the European Economics and Social Committee and the Committee of the Regions COM(2005) 35 final, Commission of the European Communities, Brussels. http://europa.eu.int/eur-lex/lex/LexUriServ/site/en/com/2005/com2005_0035en01.pdf

CEC (2005b), Winning the Battle against Global Climate Change – Background Paper, Commission Staff Working Paper, Commission of the European Communities, Brussels. http://europa.eu.int/comm/environment/climat/pdf/staff_work_paper_sec_2005_180_3.pdf

CEC (2006), Impact Assessment – Annex to the Communication from the Commission Reducing the Climate Change Impact of Aviation, Commission of the European Communities. Brussels, COM(2005) 459 final

CEU – Council of the European Union (1996) 1939th Council Meeting, Luxembourg, 25 June 1996

CEU – Council of the European Union (2004) 2632nd Council Meeting, Brussels, 20 December 2004

Chichilnisky G (2000) An axiomatic approach to choice under uncertainty with catastrophic risks. Res Energy Econ 22:221-231

Clarkson R, Deyes K (2002) Estimating the Social Cost of Carbon Emissions. The Public Enquiry Unit – HM Treasury, London, Working Paper 140

Dessus S, O'Connor D (2003) Climate Policy without Tears: CGE-based Ancillary Benefits for Chile. Env Res Econ 25,3:287–317

Dixit AK, Pindyck RS (1994) Investment under Uncertainty Princeton University Press, Princeton

Dorland C, Tol RSJ, Palutikof JP (1999) Vulnerability of the Netherlands and Northwest Europe to Storm Damage under Climate Change. Climatic Change 43:513–535

Dovers SR, Handmer JW (1995) Ignorance, the Precautionary Principle, and Sustainability. Ambio 24,2:92–97

Eismont O, Welsch H (1996) Optimal Greenhouse Gas Emissions under Various Assessments of Climate Change Ambiguity. Env Res Econ

Ekins P (1996) The Secondary Benefits of CO_2 Abatement: How Much Emission Reduction Do They Justify? Ecol Econ 16:13–24

Fankhauser S (1995) Valuing Climate Change – The Economics of the Greenhouse. EarthScan, London

Forest CE, Stone PH, Sokolov AP, Allen MR, Webster MD (2002) Quantifying Uncertainties in Climate System Properties with the Use of Recent Climate Observations. Science 295:113–117

Godal O, Fuglestvedt JS (2002) Testing 100-year global warming potentials: Impacts on compliance costs and abatement profile. Climatic Change 52:93–127

Hammitt JK (1995) Outcome and Value Uncertainty in Global-Change Policy. Climatic Change 30:125–145
Harvey F (2007) China Exploiting Kyoto Loophole. Financial Times (Jan 17). http://www.ft.com/cms/s/9bdda3d4-a676-11db-937f-0000779e2340.html
Harvey LDD (1996a) Development of a Risk-Hedging CO_2-Emission Policy, Part I: Risks of Unrestrained Emissions. Climatic Change 34:1–40
Harvey LDD (1996b) Development of a Risk-Hedging CO_2-Emission Policy, Part II: Risks Associated with Measures to Limit Emissions, Synthesis, and Conclusions Climatic Change 34:41–71
Helm D (2005) Economic Instruments and Environmental Policy. Econ Soc Rev 36,3:205-228
Holmes JK, Ellis HJ (1999) An Integrated Assessment Modeling Framework for Assessing Primary and Secondary Impacts from Carbon Dioxide Stabilization Scenarios. Env Model Assess 4,1:45–63
Hourcade, JC, Shukla PR, Cifuentes L, Davis D, Edmonds J, Fisher BS, Fortin E, Golub A, Hohmeyer O, Krupnick A, Kverndokk S, Loulou R, Richels RG, Segenovic H, Yamaji K (2001) Global, Regional, and National Costs and Ancillary Benefits of Mitigation. In: Davidson O, Metz B (eds) Climate Change 2001: Mitigation – Contribution of Working Group III to the Third Assessment Report of the Intergovernmental Panel on Climate Change. Cambridge University Press Cambridge
House of Lords (2005) The Economics of Climate Change. Authority of the House of Lords, London, HL 12-I
Kandlikar M (1996) Indices for Comparing Greenhouse Gas Emissions: Integrating Science and Economics. Energy Econ 18:265–281
Kandlikar M (1995) The Relative Role of Trace Gas Emissions in Greenhouse Abatement Policies. Energy Pol 23,10:879–883
Keller K, Bolker BM, Bradford DF (2004) Uncertain climate thresholds and optimal economic growth. J Env Econ Manag 48:723–741
Kolstad CD (1996) Learning and Stock Effects in Environmental Regulations: The Case of Greenhouse Gas Emissions. J Env Econ Manag 31:1–18
Lempert RJ, Nakicenovic N, Sarewitz D, Schlesinger ME (2004) Characterizing climate-change uncertainties for decision-makers. Climatic Change 65:1–9
Manne AS, Richels RG (2001) An alternative approach to establishing trade-offs among greenhouse gases. Nature 410:675–677
Manne AS, Richels RG (1992) Buying Greenhouse Insurance – The Economic Costs of CO_2 Emission Limits. MIT Press, Cambridge
Martin PH (1997) If You Don't Know How To Fix It, Please Stop Breaking It – The Precautionary Principle and Climate Change. Found Sci 2:263–292
Meinshausen M (2005) On the Risk of Overshooting 2 °C. International Symposium on Stabilisation of Greenhouse Gases. Exeter, February 1–3
Michaelis L (1997) Special Issues in Carbon/Energy Taxation: Carbon Charges on Aviation Fuels – Annex 1 Export Group on the United Nations Framework Convention on Climate Change Working Paper no. 12. Organi-

zation for Economic Cooperation and Development, Paris, OCDE/GD(97)78

Morgan MG, Henrion M (1990) Uncertainty – A Guide to Dealing with Uncertainty in Quantative Risk and Policy Analysis. Cambridge University Press, Cambridge

Olsthoorn AA (2001) Carbon Dioxide Emissions from International Aviation: 1950-2050. J Air Transp Management 7:87–93

Pate-Cornell E (1996) Uncertainties in Global Climate Change Estimates. Climatic Change 33:145–149

Pearce DW (2003) The social cost of carbon and its policy implications. Oxford Rev Econ Policy 19,3:1–32

Pearce DW, Achanta AN, Cline WR, Fankhauser S, Pachauri R, Tol RSJ, Vellinga P (1996) The Social Costs of Climate Change: Greenhouse Damage and the Benefits of Control. In: Bruce JP, Lee H, Haites EF (eds) Climate Change 1995: Economic and Social Dimensions of Climate Change – Contribution of Working Group III to the Second Assessment Report of the Intergovernmental Panel on Climate Change, pp. 179–224 (Chap. 6). Cambridge University Press, Cambridge

Pizer WA (1999) The optimal choice of climate change policy in the presence of uncertainty. Resource and Energy Economics 21:255–287

Pratt JW, Raiffa H, Schlaifer R (1995) Introduction to Statistical Decision Theory. MIT Press, Cambridge

Ramaswamy V, Boucher O, Haigh J, Hauglustaine D, Haywood J, Myhre G, Nakajima T, Shi GY, Solomon S (2001) Radiative Forcing of Climate Change. In: Houghton JT, Ding Y (eds) Climate Change 2001: The Scientific Basis – Contribution of Working Group I to the Third Assessment Report of the Intergovernmental Panel on Climate Change. Cambridge University Press, Cambridge, pp 349–416

Schmalensee R (1993) Comparing Greenhouse Gases for Policy Purposes. Energy J 14:245–255

Schwank O (2004) Concerns about CDM Projects Based on Decomposition of HFC-23 Emissions from 22 HCFC Production Sites. Infras Zurich. http://cdm.unfccc.int/public_inputs/inputam0001/Comment_AM0001_Schwank_081004.pdf

Shine KP, Fuglestvedt JS, Hailemariam K, Stuber N (2005) Alternatives to the global warming potential for comparing climate impacts of emissions of greenhouse gases. Climatic Change 68:281–302

Smith SJ, Wigley TML (2000a) Global Warming Potentials: 2. Accuracy. Climatic Change 44:459–469

Smith SJ, Wigley TML (2000b) Global Warming Potentials: 1. Climatic Implications of Emissions Reductions. Climatic Change 44:445–457

Stern NH, Peters S, Bakhshi V, Bowen A, Cameron C, Catovsky S, Crane D, Cruickshank S, Dietz S, Edmonson N, Garbett SL, Hamid L, Hoffman G, Ingram D, Jones B, Patmore N, Radcliffe H, Sathiyarajah R, Stock

M, Taylor C, Vernon T, Wanjie H, Zenghelis D (2006) Stern Review: The Economics of Climate Change. HM Treasury, London

Sygna L, Fuglestvedt JS, Aaheim HA (2002) The adequacy of GWPS as indicators of damage costs incurred by global warming. Mitig Adapt Strat Glob Change 7:45-62

Tol RSJ (1999) The Marginal Costs of Greenhouse Gas Emissions: Energy J 20,1:61–81

Tol RSJ (2003) Is the uncertainty about climate change too large for expected cost-benefit analysis? Climatic Change 6:265–289

Tol RSJ (2005a) The marginal damage costs of carbon dioxide emissions: an assessment of the uncertainties. Energy Pol:33:2064–2074

Tol RSJ (2005b) The Marginal Damage Costs of Carbon-dioxide Emissions. In Helm D (ed) Climate-Change Policy. Oxford University Press, Oxford

Tol RSJ (2006) The Stern Review of the Economics of Climate Change: A Comment. Energy & Environ 17,6:977–981

Tol RSJ The Impact of a Carbon Tax on International Tourism. Transportation Res D 12,2: 129–142

Tol RSJ, FitzGerald J (2007) Europe's Airline Emissions Plan is a Flight of Fancy. Financial Times (January 18)

Tol RSJ, Heintz RJ, Lammers PEM (2003) Methane emission reduction: An application of FUND. Climatic Change 57:71–98

Tol RSJ, Yohe GW (2006) A Review of the Stern Review. World Econ 7,4: 233–250

Tuinstra W, Ridder W de, Wesselink LG, Hoen A, Bollen JC, Borsboom JAM (2005) Aviation in the EU Emissions Trading Scheme – A First Step towards Reducing the Impact of Aviation on Climate Change. Netherlands Environmental Assessment Agency. Bilthoven, 500043001

Ulph A, Maddison DJ (1997) Uncertainty, Learning and International Environmental Policy Coordination. Env Res Econ 9:451–466

van Vuuren DP, Weyant JP, Chesnaye FC de la (2006) Multi-gas Scenarios to Stabilize Radiative Forcing. Energy Econ 28:102–120

van Vuuren DP, Cofala J, Eerens HE, Oostenrijk R, Heyes C, Klimont Z, Elzen MGJ den, Amann M (2006) Exploring the Ancillary Benefits of the Kyoto Protocol for Air Pollution in Europe. Energy Pol 34,4:444–460

WBGU (1995) Szenario zur Ableitung CO_2-Reduktionsziele und Umsetzungsstrategien – Stellungnahme zur ersten Vertragsstaatenkonferenz der Klimarahmenkonvention in Berlin (Scenario to Deduct CO_2 Reduction Targets und Implementation Strategies – Position for the First Conference of the Parties of the Framework Convention on Climate in Berlin). Wissenschaftlicher Beirat der Bundesregierung Globale Umweltveränderungen, Dortmund

WBGU (2003) Über Kioto hinaus denken – Klimaschutzstrategien für das 21. Jahrhundert (Thinking beyond Kyoto – Climate Protection Strategies for the 21st Century), Wissenschaftlicher Beirat der Bundesregierung Globale Umweltveränderungen. Berlin

Webster M, Forest C, Reilly J, Babiker M, Kicklighter DW, Mayer M, Prinn R, Sarofim M, Sokolov A, Stone P, Wang C (2003) Uncertainty analysis of climate change and policy response. Climatic Change 61:295–320

Wit RCN, Dings, Mendes de Leon P, Thwaites L, Peeters P, Greenwood D, Doganis R (2002) Economic Incentives to Mitigate Greenhouse Gas Emissions from Air Transport in Europe, CE Delft, 02.4733.10

Wit RCN, Boon BH, van Velzen A, Cames A, Deuber O, Lee DS (2005) Giving Wings to Emissions Trading – Inclusion of Aviation under the European Trading System (ETS): Design and Impacts, CE Delft, 05.7789.20

Working Group on Public Health and Fossil-Fuel Combustion (1997) Short-Term Improvements in Public Health from Global-Climate Policies of Fossil-Fuel Combustion: An Interim Report. Lancet 350,9088:1341-1348

Yohe GW (1997) Uncertainty, Short Term Hedging and the Tolerable Window Approach. Glob Env Change (forthcoming)

Yohe GW (2006) Some Thoughts on the Damage Estimates Presented in the Stern Review – An Editorial. Integr Assess J:6,3:65–72

Zapert R, Gaertner PS, Filar JA (1998) Uncertainty Propagation within an Integrated Model of Climate Change. Energy Econ 20:571–598

Curbing the Omnipresence of Lead in the European Environment Since the 1970s – a Successful Example of Efficient Environmental Policy*

Hans von Storch

GKSS Research Centre, Max-Planck-Straße 1, 21502 Geesthacht, Germany

1 Introduction

For the foreseeable future, the atmosphere and the environment in general will remain to serve as a dump for various anthropogenic substances. Some substances will have negative properties so that society will sooner or later begin regulating their emissions. To that end, science must provide society with the tools for the retrospective evaluation of the physical and economical impacts of past regulations, and for the predictive evaluation of alternative scenarios of future regulations.

We have developed such a tool for reconstructing past lead air concentrations and depositions across Europe (1958–1995), made up of a detailed emissions, a regionalized history of weather events (with the help of a regional climate model using global weather re-analyses as input), and an atmospheric transport model (for a summary, refer to von Storch et al., 2002, 2003).

We used this tool in conjunction with lead measurements in biota and human blood, and with an economic analysis to assess past European gasoline-lead regulations. Some of the specific questions asked were: How did lead emissions, atmospheric concentrations and depositions develop since the 1950s? Was the decline in air concentrations matched by corresponding declines in plants, animals and humans? Did the regulations result in considerable economic burdens in Germany? How was the media coverage of the issue of lead in gasoline?

We have chosen lead for several reasons. Lead, specifically tetraethyllead has been used for a long time as an anti-knocking additive in gasoline (cf., Berwick, 1987; Seyferth, 2003) The use of lead in gasoline underwent significant changes, from an unabated increase of emissions to a series of sometimes

* This article is an updated and extended version of: von Storch, H., C. Hagner, M. Costa-Cabral, F. Feser, J. Pacyna, E. Pacyna and S. Kolb (2002) Reassessing past European gasoline lead policies. EOS 83, p. 393+399

drastic reductions of emissions since the 1970s. Thus, there is a strong and well-defined signal to be detected. Second, once released into the atmosphere, lead will accumulate and persist almost indefinitely in some environmental compartments, such as aquatic sediments. What might the ecological and human health impacts be when this neurotoxin is present in the environment? Finally, airborne lead behaves to a first order approximation as inert, so that the simulation of its transport and deposition is relatively simple. In principle, our tool can be used for any other particle-bound substance of limited reactivity.

It turns out that this approach is successful in describing the temporal evolution of the spatial distribution of lead deposition in Europe. Demonstrating the effectiveness of gasoline-lead policies, the reconstructed concentrations in the atmosphere, in plant leaves and in human blood show a steady decline since the early 1980s, while concentrations in marine organisms along the North Sea coast, however, seem to remain unaffected – at least until recently. Contrary to initial expectations, the German mineral oil industry was not negatively affected. While competition conditions changed in the German gasoline and automobile markets, no impacts of the regulations could be identified in the macro-economic indicators. While lead pollution has successfully been curbed in Europe and North America, the problem persists in many parts of the world, including Africa.

2 Gasoline-Lead Regulations in Europe

Air pollution problems related to automobile traffic in the 1960s were addressed in the US by the 1963 Clean Air Act.[2]. In Europe, concern with the resulting risks to human health would only gain momentum in the 1970s. Lead in particular, which was added to gasoline for its anti-knocking properties, was perceived as a health threat at this time, given the then new evidence of its severe neurotoxicological effects, especially to children. After lead-based paint and lead solder in water pipes and food cans was prohibited, gasoline lead (tetraethyl and tetramethyl lead additives) became the next target.

In the 1970s, the German government was the first in Europe to regulate gasoline lead. A maximum content of 0.4 g Pb/l was imposed in Germany in 1972 (down from the usual 0.6 g Pb/l) and lowered further to 0.15 g Pb/l in 1976. A preliminary analysis of newspaper coverage found that the topic of gasoline lead health dangers entered the German press in the 1960s. British articles did not focus on lead but on urban smog instead. Also, in 1972, a group of experts from the French government did not acknowledge any automobile emissions to be dangerous (Kolb, 2005). Starting only in 1981, the

[2] Actually, the debate about the health implications of using lead in gasoline began already in the United States in the 1920s. An account of this exciting and sometimes harrowing history is provided by Kitmann (2000). See also Seyferth (2003).

European Union (EU) fixed its limit modestly at 0.4 g Pb/l (Council Directive 78/611/EEC of 1978) (Hagner, 2000).

In the 1980s, the discussion of automobile air pollution in Europe moved to concerns relating to forest protection and the effects on forests of massive NO_x, CO, and CxHy emissions. This discussion was also initiated by Germany, concerned with the death of the 'German Forest' from acid rain and photo-oxidation (Kolb, 2005).[3] In 1985 Germany passed a law to reduce total automobile emissions. This law included the introduction of unleaded gasoline since the largest reductions in NO_x, CO, CxHy and other pollutants could only be achieved with catalytic converters (already in use in the US) and these were incompatible with lead. Opposing reactions expressed in the media of some European countries are reviewed by Kolb (2005; for a short English account refer to von Storch et al. (2003). The press coverage in the 1980s emphasized the expected economic problems in the automobile industry and the difficulty in finding a compromise solution in Europe.

A detailed account of the socio-political process which led to the regulation of lead use in gasoline in Switzerland is given by Breu et al. (2002) and Mosiman et al. (2002). A pan-European account of the introduction of unleaded gasoline is provided by Berwick (1987). Despite the opposition, in 1985 the EU mandated all member states to offer unleaded gasoline starting October 1989, and recommended a maximum of 0.15 g Pb/l. While some countries promptly adhered to this directive, others lagged behind (see Hagner, 2000). The Aarhus Treaty, signed in 1998 by nearly all European countries, stipulated the exclusive use of unleaded gasoline in automobiles by the year 2005.

3 Reconstructing Regional Pathways of Lead

For running a model of atmospheric lead transport, regional weather information including wind speed and direction, precipitation rate and boundary layer depth, are required. Since the global weather analyses are available from NCEP (Kalnay et al., 1996) since 1958 at 2° spatial resolution were considered too coarse, the regional atmospheric model REMO was used to "downscale" to a 0.5° grid (roughly, 50 km scale) covering all of Western Europe and parts of the North Atlantic (Feser et al., 2001).

Emission estimates disaggregated to the 0.5° grid were provided by Pacyna and Pacyna (2000) for the years 1955, 1965, 1975, 1985, 1990 and 1995. Figure 1 shows the yearly totals, peaking at nearly 160,000 tons in 1975, and shows the predominance of automobile emissions. The sharp decrease since the 1970s resulted from the gasoline-lead regulations as well as from the abatement of fixed-source lead emissions (industrial and others).

[3] The media coverage of the legal efforts to reduce the use of lead as an antiknocking additive in gasoline is covered by Steffen Kolb in another contribution to this volume.

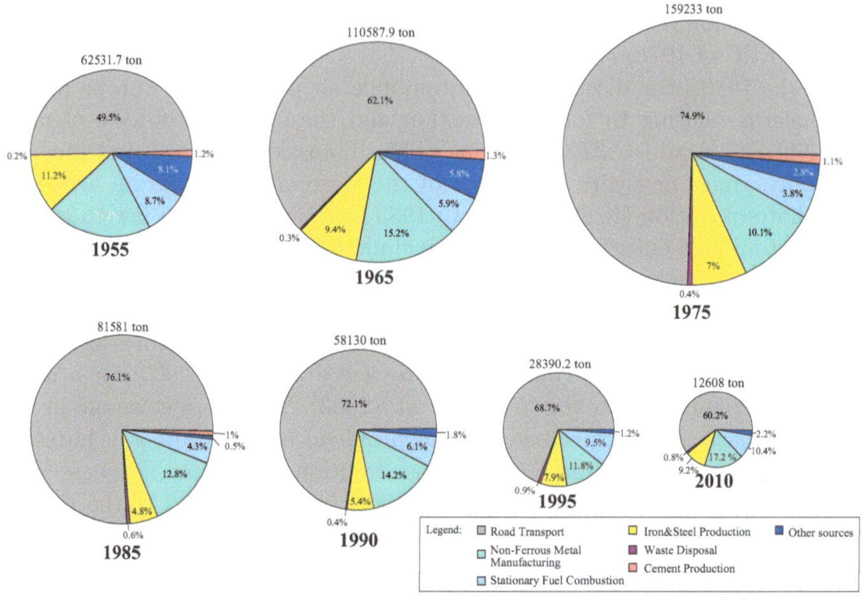

Fig. 1. European lead emissions estimates by source category (from Pacyna and Pacyna, 2000)

Using these emission estimates and the regionalized atmospheric forcing, lead concentrations and depositions over Europe were computed using a two-dimensional Lagrangian model (Costa-Cabral, 1999), with a 6-hourly time step and a 0.5° spatial resolution. It was assumed that lead remains within the well-mixed planetary boundary layer, where it is horizontally advected by wind and deposited to the surface by turbulent transport and precipitation scavenging. The dry settling velocity used was $0.2\,\mathrm{cm\,s^{-1}}$ and the precipitation scavenging constant was 5×10^5.

To validate the model results, a comparison was made with local measurements of lead concentrations and depositions obtained from EMEP (for details refer to von Storch et al., 2003). The general pattern of deposition since the beginning of monitoring activities in 1960 are reproduced very well by the model, particularly those of the early 1980s. The added value provided by the model is the complete space-time coverage, extending over two decades (1960–1980) while using hardly any observations.

Simulation results indicate that most of the deposition within a given country originates from its own emissions. Only smaller countries such as Switzerland or the Netherlands have suffered from substantial depositions originating in neighboring states (von Storch et al., 2003). For the Baltic Sea, for instance, 23% of the total depositions originate in Poland, 20% in Germany, and 16% in Finland. According to our estimation, total input peaked in the mid-1970s, surpassing 3.500 tons annually, and declined to under 500 tons in

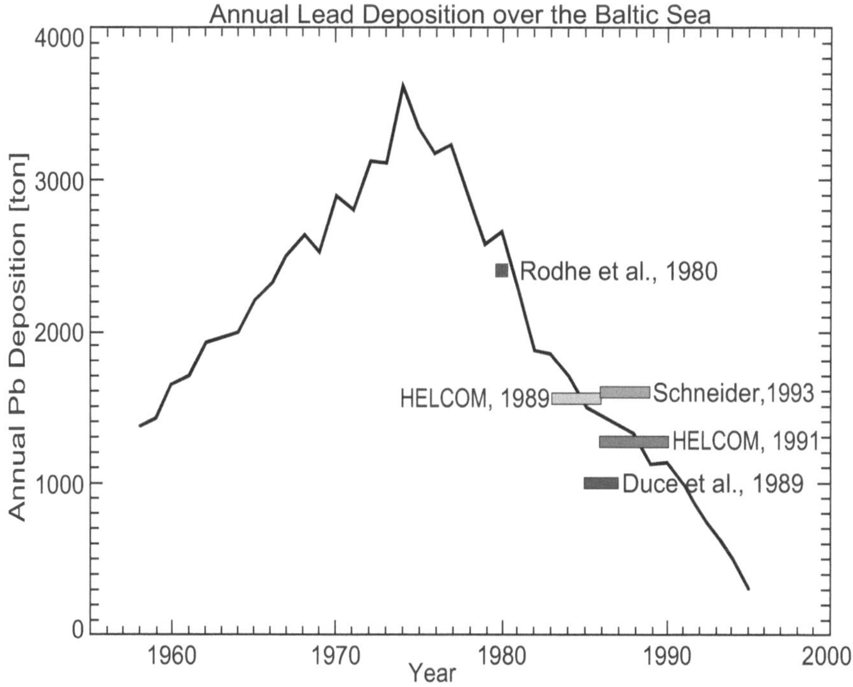

Fig. 2. Simulated input of lead into the Baltic Sea (*line*) and estimates based on comprehensive analyses of observational data (*colored bars*) (von Storch et al., 2003)

1995 (Fig. 2). Simulations compare favorably with comprehensive analyses of the overall deposition of lead into the Baltic Sea based on observational evidence obtained during the second half of the 1980s (Fig. 2).

Schulte-Rentrop et al. (2005) extended the analysis – by simulating the overland transport of lead after it had been deposited at the earths surface. They considered the catchment area of the river Elbe and described the transport of lead into the river by way of atmospheric deposition onto the surface of the river, by erosion and by runoff. A major finding of this study was that the flux of lead into the river diminished since the 1970s, but that the ongoing deposition, with decreasing rates since the 1970s, was associated with a steady accumulation of lead in soils (see also Johansson et al., 2001). Accordingly, the soil released (via erosion and runoff) increasing amounts of lead into the river. Only the atmospheric depositions into the river decreased in parallel to the decreasing atmospheric loads. The overall effect was, however, a reduction in the flux of lead into the river (Fig. 3).

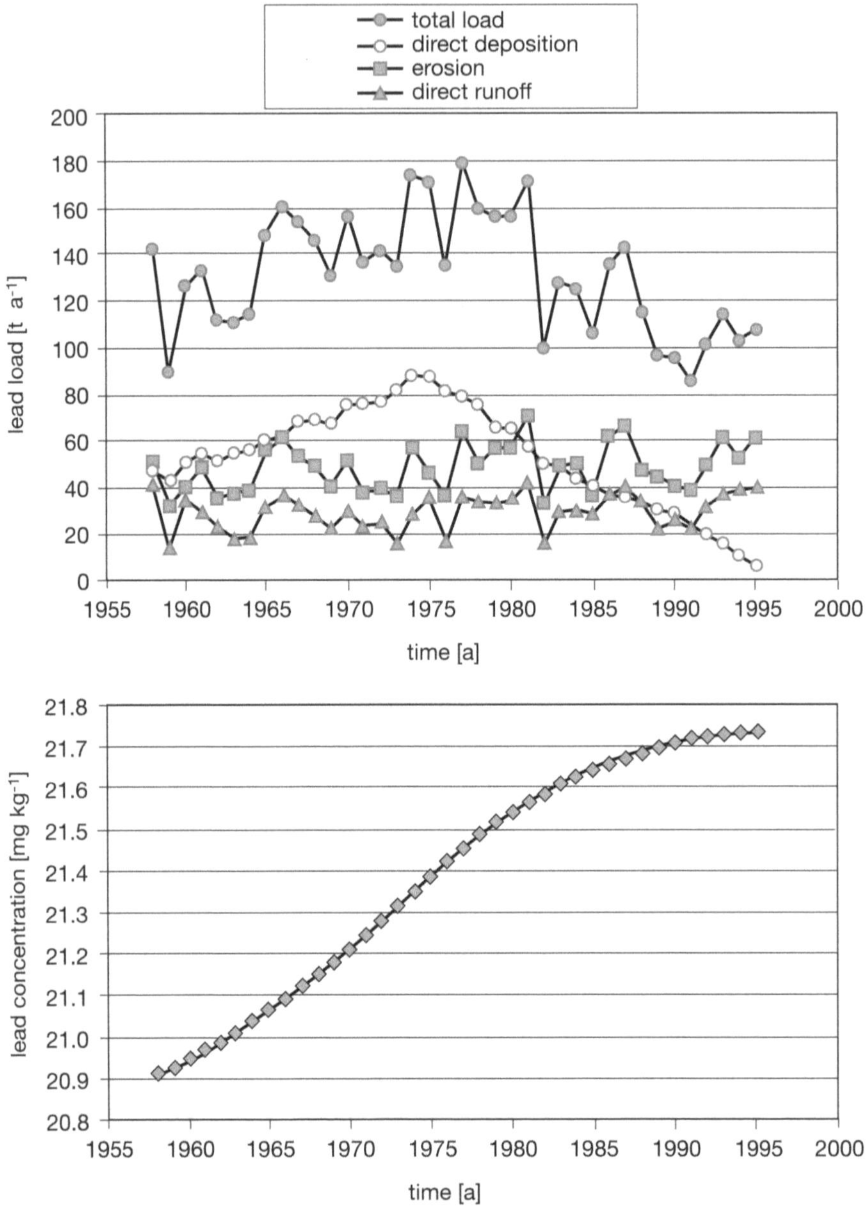

Fig. 3. *Top:* Simulated lead loads via the three pathways into the river – direct deposition, erosion and run-off. *Bottom:* Mean simulated lead concentration of the soils in the Elbe catchment

4 Some Environmental and Economical Impacts

Measurements made in Germany between the 1980s and 1990s showed that atmospheric lead concentrations halved approximately every 4.5 years (Hagner, 2000). The same trend could be observed in plants, for example, between 1985–1996 a decline in lead concentrations was observed in annual spruce needles and poplar leaves in Germany. However, in marine organisms, such as, for example, the blue mussel and fish in the German Wadden Sea, lead levels have not diminished since the 1980s (Fig. 4; Hagner, 2001).

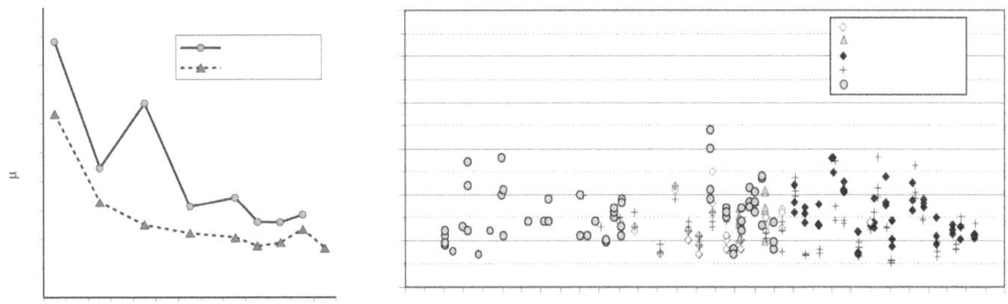

Fig. 4. *Left:* Lead concentrations (mg/g) in spruce (Picea abies) sprouts and poplar (Populus nigra) leaves in urban areas in Saarland (Germany), 1985–1996. *Right:* Lead concentration (mg/kg) in Blue Mussels (Mytilus Edulis) along the southern in North Sea coast, 1982–1997. After Hagner (2002)

Between 1979 and 1997, several studies investigated the levels of lead in human blood in Germany (see Hagner, 2002). During this period, levels remained consistently below those indicated as hazardous to adults by medical experts. In Fig. 5, blood lead levels are crudely estimated back to the year 1958. To do this, a regression-type model was constructed using the recorded lead concentrations in adult human blood and the simulated aerial lead concentrations in one grid box (von Storch and Hagner, 2004). Data on sample mean concentrations (i.e., mean values across a sample of many people) as well as the 90 and 95%iles were available, so that it could be concluded that 10% (or 5%) of the people sampled had lead concentrations above the 90%ile (95%ile) in their blood, and 90% (95%) had levels below this number. In Fig. 5, the colored backgrounds indicate the critical levels as stipulated by the German Human Biomonitoring Commission. For levels above 250 µg Pb/l health risks for adults are expected, and for levels above 150 µg Pb/l monitoring is advised (Hagner, 2000). For pregnant women and for children, a critical value of 150 µg Pb/l was adopted as indicative of a potential risk to health [4].

[4] Interestingly, some American researchers believe that the intellectual development of children is already disturbed at a blood lead level of 100 µg Pb/l.

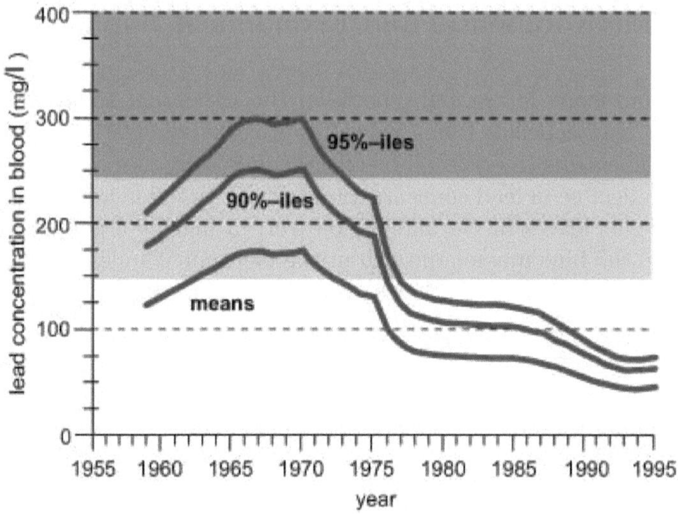

Fig. 5. Estimated lead concentration in adult human blood in Germany. The lowest curve relates to the population mean, so that half of the adults have blood concentrations below the cure, and half above. The uppermost curve refers to the 95%ile, so that 5% of all adults have concentrations above the curve. The middle curve describes the 90%ile. The three coloured backgrounds refer to the domains used by the German Human Biomonitoring Commission to classify health risks – in the white domain no dangers are expected, and in the upper class (*densely stippled*) a health risk prevails (von Storch and Hagner, 2004)

The estimated mean blood levels of lead reached a peak level of about 150 μg Pb/l in the early 1970s (von Storch and Hagner, 2004). This implies that it is highly likely that the lead concentrations in the ambient air in the mid-1970s may have been high enough to raise serious medical concerns for half of the population. The 90%ile reached a level of 250 μg Pb/l, and the 95%ile reached a level of 300 μg Pb/l. Thus, 10% of the entire population was exposed to a serious health risk in the early 1970s [5]. After this period the levels diminished and have now reached much lower levels, well below the critical levels recommended by the Human Biomonitoring Commission. For an international comparison, refer to Thomas et al. (1999).

An assessment of the most immediate economical impacts of the lead regulations is a difficult task. Hagner's (2000) analysis indicates that despite the concerns voiced by the German mineral oil industry that petrol production would become costlier following the first regulation in 1972, the costs have actually dropped thanks to the savings resulting from lead additives. It was only after the second regulation in 1976 that production costs rose slightly

[5] At least for the site where the blood monitoring was carried out, namely the city of Münster in Nordrhein-Westfalen.

and this was due to the fact that new additives with high octane numbers were now required for maintaining gasoline performance (Hagner, 2000).

The impacts of introducing unleaded gasoline in 1985 were more complex. Tax incentives for unleaded gasoline and for low-emission cars increased the sales of both products. Many independent gasoline traders went bankrupt, as gas-station reconstruction represented a higher financial strain for them compared to the large multinational companies. Favorable terms of competition were experienced by car manufacturers with the highest technical standards, and who had already gathered experience with catalyst systems on the U.S. market (Hagner, 2000).

Aside from these shifts in market competition conditions, no significant impact could be detected in the German macro-economic indicators including gross national product, economic growth, price stability, unemployment level, or foreign trade balance.

5 Conclusion and Outlook

We have developed a tool for reconstructing past lead air concentrations and depositions across Europe. With the help of regionalized atmospheric data, spatially disaggregated lead emissions from road traffic and point sources, and various local data, an attempt was made to reconstruct the airborne pathways and deposition of gasoline lead in Europe since 1958. We have also analyzed trends in concentrations in biota and human blood, and evaluated the most direct economic impacts of gasoline-lead regulations.

We have demonstrated that for the case of lead our tool is functioning well. Our modelled data show that European gasoline-lead reduction regulations may be considered a good example of successful environmental policy. However, the success of lead policies is limited to atmospheric pathways, and did not have the effect of lowering concentrations in some marine biota and soil concentrations.

One should, however, not forget that the large amounts of lead emitted in the past 50 years have not simply vanished but ubiquitously reside for good in the global environment. The use of lead in gasoline was indeed a large-scale geophysical pollution exercise, and it remains to be seen if long-term effects may emerge at a later time.

The major conclusion to be drawn from our analysis is that the regulation to remove lead as an anti-knocking additive in gasoline after two decades of unabated increase was mostly successful. The regulation solved the problem for the atmospheric pathway, and that part of the ecosystem in which the toxin essentially "passes through" in relatively short time. Thus, a relatively short residence time is a necessary condition for substance abatement through emission regulations in a given environmental compartment once considerable amounts of the substance have already been released. For other parts of the ecosystem, which accumulate and store the toxin for an extended period of

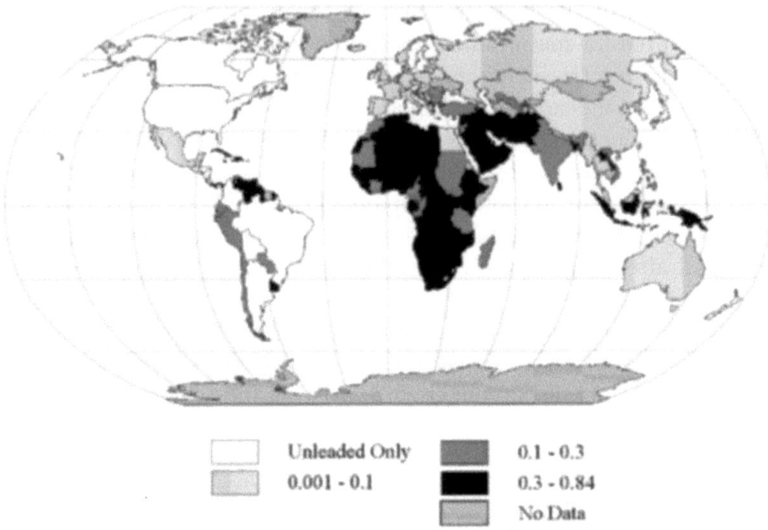

Fig. 6. Use of leaded gasoline in different countries of the world. After Thomas and Kwong (2001). The unit is g Pb/l. Values in most of Europe are 0.0015 g Pb/l – which is sold as "lead-free"

time, such a regulation is better than nothing but is certainly insufficient. For such systems, the prevention of such pollution actually occuring is the only solution.

The success of the environmental protection policy was mostly limited to the developed world (Fig. 6, Thomas, 1995, Thomas et al., 1999; Thomas and Kwong, 2001). Almost everywhere in Africa and the Near East, where lead was still in use as an anti-knocking additive up until the early 2000s and on a large scale, the lead concentrations were mostly 0.3 g Pb/l and higher. For instance, Thomas et al. (1999) report about 0.39 g Pb/l in gasoline and 156 µg Pb/l in human blood for Caracas in 1991; in Mexico City the concentration went down from 0.2 g Pb/l in gasoline and 122 µg Pb/l in human blood in 1988 to 0,06 g Pb/l in gasoline and 70 µg Pb/l in human blood in 1993. For Cape Town the latest concentrations available are from 1990 with 0.4 g Pb/l in gasoline and 72 µg Pb/l in human blood. In Egypt all gasoline had lead in it in 1994, with 0.35 g Pb/l (Thomas, 1995); the same account reports 0.66 g Pb/l in Nigeria (no year given) and 0.84 g Pb/l in Equador (in 1993). One would hope that the situation has improved in these countries.

Our methodology is presently being extended to more "interesting" chemicals, in particular to benzo(a)pyrene (Matthias et al., 2007; Aulinger et al., 2007).

Acknowledgements

I am grateful to the "lead group" at GKSS: Charlotte Hagner, Mariza Costa-Cabral, Frauke Feser, Annette-Schulte-Rentrop; to Jozef Pacyna and Elisabeth Pacyna, who constructed the emission maps; to Steffen Kolb, who analysed the contemporary media discourse.

For further information refer to: http://guasun2.gkss.de/cgi-bin/datawwwiz?db=lead The annual emissions, and modeled concentrations and depositions data are available for download from a link on this page.

References

Aulinger A, Matthias V, Quante V, Quante M (2007) Introducing a partitioning mechanism for PAHs into the Community Multiscale Air Quality modelling system and its application to simulating the transport of benzo(a)pyrene over Europe. J Appl Meteorol, submitted

Berwick I (1987) The rise and fall of lead in petrol. Phys technol 18:158-164

Breu M, Gerber S, Mosimann M, Vysusil T (2002) Bleibenzin – eine schwere Geschichte. Die Geschichte der Benzinverbleiung aus der Sicht der Politik, des Rechts, der Wirtschaft und der Ökologie. Ökom Verlag, München

Costa-Cabral MC (1999) TUBES: An exact solution to advective transport of trace species in a two-dimensional discretized flow field using flow tubes. GKSS report GKSS 99/E/60

Feser F, Weisse R, von Storch H (2001) Multi-Decadal Atmospheric Modeling for Europe Yields Multi-purpose Data. EOS, American Geophysical Union, 82,28:305–310

Hagner C (2000) European regulations to reduce lead emissions from automobiles – did they have an economic impact on the German gasoline and automobile markets? Reg Env Change 1:135–151

Hagner C (2002) Regional and Long-Term Patterns of Lead Concentrations in Riverine, Marine and Terrestrial Systems and Humans in Northwest Europe. Water Air Soil Poll 134:1–40

Johansson K, Bergbäck B, Tyler G (2001) Impact of atmospheric long range transport of lead, mercury and cadmium on the Swedish forest environment. Water Air Soil Poll: Focus 1:279–297

Kalnay E, Kanamitsu M, Kistler R, Collins W, Deaven D, Gandin L, Iredell M, Saha S, White G, Woollen J, Zhu Y, Chelliah M, Ebisuzaki W, Higgins W, Janowiak J, Mo KC, Ropelewski C, Wang J, Leetmaa A, Reynolds R, Jenne R, Joseph D (1996) The NCEP/NCAR 40-Year Reanalysis Project. Bull Am Met Soc 77,3:437-471

Kitmann JL (2000) The secret story of lead. The Nation 270,11:11-44

Kolb S (2005) Mediale Thematisierung in Zyklen. Theoretischer Entwurf und empirische Anwendung. Dissertation Universität Hamburg. Herbert von Halem Verlag, Köln

Matthias V, Aulinger A, Quante M (2007) Adapting CMAQ to investigate air pollution in North Sea coastal regions. Env Modelling and Software (submitted)

Mosimann M, Breu M, Vysusil T, Gerber S (2002) Vom Tiger im Tank – Die Geschichte des Bleibenzins. Gaia 11:203–212

Pacyna JM, Pacyna EG (2000) Atmospheric Emissions of Anthropogenic Lead in Europe: Improvements, Updates, Historical Data and Projections. GKSS Report 2000/31, GKSS Research Center

Schulte-Rentrop A, Costa-Cabral M, Vink R (2005) Modelling the overland transport of lead deposited from the atmosphere in the Elbe catchment over four decades (1958–1995). Water, Air and Soil Poll 160,1–4:271-291

Seyferth, D (2003) The rise and fall of tetraethyllead. 2. Organometallics 22:5154-5178

Thomas V (1995) The elimination of lead in gasoline. An Rev Energy Env 20:301–324

Thomas VM, Socolow RH, Fanelli JJ, Sprio TG (1999) Effects of reducing lead in gasoline: an analysis of the international experience. Env Sci Technol 33:3942–3947

Kwong Thomas, Kwong A (2001) Ethanol as a Lead Replacement: Phasing Out Leaded Gasoline in Africa. Energy Pol 29:1133-1143

von Storch H, Costa-Cabral M, Hagner C, Feser F, Pacyna J, Pacyna E, Kolb S (2003) Four decades of gasoline lead emissions and control policies in Europe: A retrospective assessment. Sci Tot Env (STOTEN) 311:151–176

von Storch H, Hagner C, Costa-Cabral M, Feser F, Pacyna J, Pacyna E, Kolb S (2002) Reassessing past European gasoline lead policies. EOS, American Geophysical Union 83:393+399

von Storch H, Hagner C (2004) Controlling lead concentrations in human blood by regulating the use of lead in gasoline. A case study for Germany. Ambio 33:126–132

Mission Impossible? Media Coverage of Scientific Findings

Steffen Kolb[1] and Steffen Burkhardt[2]

[1] Hamburg Media School, Finkenau 35, 22081 Hamburg, Germany,
[2] Hamburg Media School, Finkenau 35, 22081 Hamburg, Germany

1 Introduction

At a first glance, journalistic coverage of research results from the world of academia seems to be possible without any problems because scientific data is generated for the purpose of publishing. However, scientific publications aim at a specific target group often called the "scientific community". In contrast, journalistic publications – at least those produced for news media – aim at, in most cases, the general public. These different target groups have different habits of consuming information and have different expectations on how to be addressed. Whereas the general public looks for a simple explanation of facts, the scientific community expects an in-depth discussion of specific problems.

The resulting differences in the demands on scientific versus journalistic information lead to the central question of this paper: Is journalistic work based on scientific findings a mission impossible or is there a way of combining complexity and simplicity in one text?

To answer this question, this paper first specifies the aims and functions of journalism versus science in Western democratic societies. Second, it compares the two systems by working out what the main differences are from among the numerous similarities between journalistic and scientific publications. To scrutinise the way in which journalists "tell their stories", a theory of issue life-cycles developed from the social and life sciences' findings is presented in the third part of the paper. This theory allows for an analysis of the life-cycle of issues by identifying typical phases of media coverage. As a case study, the fourth chapter applies this theory to the British, French, and German media coverage of the "lead problem"[3] as a case study. This paper comes to the conclusion that media coverage of scientific data tends to focus on the

[3] Since the late 1960s the poisoning effects of tetra-ethyl-lead (TEL), which had been introduced into car fuel as a component to reduce engine knocking, are well known.

most dramatic findings with the most social impact. If researchers do not also present the social relevance of their findings, then these will often be ignored.

2 Journalism and Science: Differences and Similarities in Aims and Functions of two Systems

At first sight, it seems rather absurd to compare the two systems of science and journalism because their aims and functions within society are obviously rather different. However, the scientific methodology of comparative research allows the comparison of everything that can be put on the same conceptual basis and which are not equal: "[T]he comparison of equal or identical elements would offer little results other than the equality of these elements (van Deth, 1998). The Anglo-American metaphor of comparing apples and oranges is as misleading as the common use of the term 'comparable' for describing two or more similar objects. Apples and oranges are, in fact, quite comparable, due to their common functional integration into the 'concept' of fruit (Aarebrot and Bakka, 1997)" (Wirth and Kolb, 2004). In our case, the similar conceptual basis for science and journalism can be assumed due to the integration of science and journalism in the same social context: Any of the nations under investigation in the empirical part of this paper (United Kingdom, France, and Germany) maintain a journalistic as well as a scientific system.

The aim of any comparison has to be the search for similarities and differences. Based on Mill's and Przeworski/Teune's approach the search for differences is theoretically most interesting whenever the systems under investigation are rather similar (Mill, 1843,1959; Przeworksi and Teune, 1970). However, one could also think that the best results can be achieved for comparisons when the systems are rather different. To combine these basic assumptions one can also compare systems in an iterative process. This is usually leading to a broad understanding regardless of the degree to which the objects in question resemble each other (Kolb, 2004). Starting from the postulation that journalism and science are rather different systems within a given society, one would start the iteration process by attempting to find similarities. This is done in the following section of this paper.

There are many differences between journalism and science concerning, for example, public or private organisation, and education of the people involved. However, in both disciplines the basic method of working is called "research", and this is no coincidence. The term research describes the search for information. Both journalists and scientists do research to gain and accumulate knowledge about certain facts. However, the methods of research which are used to gain this knowledge do vary. Not only do they differ between science and journalism but they also differ within different scientific disciplines and different journalistic cultures. For the production of knowledge, journalism and science use specific professional research techniques such as, for example, monitoring, surveys and interviews, although the standards and spectrum of the methods used in scientific research are more complex than those employed in journalism.

Moreover, research in both systems can generally be described as a systematic operation of professional inquiry aimed at discovering, interpreting, and revising facts. This analytical investigation describes a collection and accumulation of facts that produces a better understanding of events, situations or actions in a social system or its environment. Both journalism and science gain and communicate knowledge and generate cognition by researching and providing information for society Rühl, 1980). Research results are the starting point for the production of meaning in science and journalism. Therefore, they are at the same time a basis for the construction of meaning in society. While scientific research is organised systematically, institutionalised research methods in journalism have, in the most cases, more of an explorative and subjective character.

However, in spite of sharing a lot of the main ideas of research, there are four important differences between journalistic and scientific research as Weischenberg has pointed out (Weischenberg, 1998): First of all, science generally works out regularities and orderliness whereas journalism searches extraordinary situations, processes or actions. “News is what differs” (Weischenberg, 1998). Second, journalism focuses on the discovery of current problems, whereas science concentrates above all on the solution of long-term problems. Third, the economical and organizational resources of journalism are always more limited than the resources of science. Fourth, because of these main differences between journalism and science, the presented concepts of reality have different functions and aims.

The functions and aims of science depend on the principals of a research project. Fundamental research, which can be called “pure research” because of its great independence from principals, has the advancement of knowledge and the theoretical understanding of the relationships beween variables as its main objective. It is exploratory and conducted without any practical end in mind, although it may have unexpected results pointing to practical applications. Fundamental research provides the basics for the further generation of theory (Krotz, 2005).

In contrast to fundamental research, applied research is carried out to solve specific, practical questions. The primary aim of this application-oriented science is not to gain knowledge or find truth for its own sake, but to explain reality. Therefore, specific scientific methods are fundamental to the investigation and acquisition of knowledge about social systems and their environments. Scientists use observations, hypotheses, and logic to propose explanations for natural phenomena. In the form of hypotheses and predictions these theories can be reproducibly tested by experiment in contrast to the predictions of journalism. The main function of journalism is not to find new knowledge about reality. Journalists want to tell their audiences an objective story on how the world is; they do so by collecting and selecting actual themes from diverse contexts Weischenberg et al., 1996).

Both journalism and science present research results by considering the idea of objectivity. In journalism, the "neutral point of view" is more an ideal schema of coverage, since certain subjective bias cannot completely be avoided. Therefore, objectivity has always been a major issue in journalism, especially in discussions about yellow journalism (Burkhard, 2005). Journalistic news professionalism in Western democracies therefore tries to avoid explicit positioning for or against a particular group or interest. Journalists are expected to publish facts without adding subjective interpretations. However, the distinction between facts and interpretations and therefore the possibility of achieving objective coverage has often been questioned.

Because of these disputes, Gaye Tuchman describes objectivity in journalism as a specific method of writing (or telling a story) which is based on five strategies (Tuchman, 1972 and 1978): the presentation of opposing point of views and possibilities of reality; the presentation of facts that support the statements; the simulation of objectivity by selecting and marking quotation; the emphasis by structuring the news and the pseudo-objective segregation of news and opinion. These strategies help to construct the reality of mass media' and journalism (Luhmann, 2004).

In science, objectivity is usually considered as the result of the application of standardised research methods by the scientific community. However, it is based on debates and agreements on certain paradigms. In history, objectivity is to be achieved by using the historical method, defined in the late 19th century and by peers review. Objectivity in science means an intersubjectively applicable, operationalised method of knowledge production which results from a discourse system with discourse rules (Schmidt, 2003). In contrast to journalism, the aim of this discourse rules is not to present a closed story but to provide an impulse for sequential communication, in order to generate objectivity by the 'self-imprinting' of the science system (Luhmann, 1990).

The review by peers as a strategy of objectifying in science, contrasts with the mostly unprofessional audience journalism corresponds with and depends on. While scientific publishing describes a system which is necessary for academic professionals in order to review work and make it available for a wider audience, media coverage is usually organised by an editorial office with a chief editor who has to maximize the profits of the media company. Traditionally, the academic system varies widely between fields, and it is also always changing, often slowly, as opposed to the high-dynamically developing agenda setting of journalistic news. Most scientific work is published in the form of articles in scientific journals or in book form. The most established academic fields have their own scientific journals, but interdisciplinary journals also exist which publish papers from several distinct fields or subfields.

The kinds of publications which are accepted as contributions to scientific findings vary greatly between fields. Therefore, the audience of scientific publications is a public of professionals which is familiar with the specific academic terms, the main concepts and ideas of its disciplines, the ethical problems, and the research conditions. Since the linguistic turn' especially, there is a specific consciousness of truth' as a construct in the scientific community, particularly in the social sciences, and this differs from journalism which sells stories as truths (Rorty, 1967; Sandbothe, 2000).

Due to the different natures of journalism, which, on the one hand, may be characterised as "immediate history", and science which is more "durable history" on the other hand, the type of objectivity welcome in both disciplines differs significantly. Journalists often use quotations of, for example, politicians' or laymen's comments on various events, situations or actions, while scientists dedicate themselves to describing the chain of causality.

Objectivity in journalism is obtained by, for example, cross-checking sources, finding primary sources and never trusting only one authority. However, the journalistic ideal of objectivity has often been criticised as impossible to achieve. Leaders of other journalistic concepts such as precision journalism, interpretative journalism, new journalism and investigative journalism see the 'truth' more as an impact of cumulative objectivities (Meyer, 1991; MagDougall and Reid, 1987; Wolfe, 1975; Weber, 1974; Mott, 1959). To sum up, all journalistic concepts have the presentation of truth as their main aim, while science wants to find out how social systems and their environments can be understood and explained.

From the point of view of both science and journalism the possibility of a complete objectivity has often been discussed. It has not only been considered as the result of a specific method but also, as in the classic Marxist conception, as the result of social interactions (Laclau and Mouffe, 1985). In that sense, the objectivity is also the result of social interactions, and even the scientific discourse cannot be disassociated from the social context.

Jürgen Habermas states that objectivity was achieved through a continuous dialogue, which could be isolated from power relations and could only lead towards further improvement and accuracy (Habermas, 1990/1962). In this concept, objectivity in the discourse is finally reached through a consensus considered as the condition of possibility of the discourse itself. According to this conception, objectivity in the public sphere requires communication and good faith. Even if one does not accept the existence of independent propositions or timeless truths, this does not exclude the possibility of viable communication or knowledge. This optimistic view of necessary progress through conversation was criticised by philosophers such as Michel Foucault (Foucault, 1974 and 1978). Despite all differences between both systems, there is a specific rapprochement from science to journalism. The way of publishing scientific research results is undergoing major changes, emerging from the transition from analogue to digital format. This change leads to a gradual mixture of science and journalism: Since about the early 1990s a major trend, particularly with respect to science journals, is open access. There are two main forms of open access: "open access publishing", in which the articles or the whole journal is freely available from the time of publication, and "self-archiving", where the authors make a copy of their own works freely available on the web.

These new forms of open access are also established in journalism. In the virtual spaces of the internet there are new hybrids of scientific oriented journalism and science presented in a journalistic way. The problems and chances of these virtual publication hybrids are not new: they follow specific implications in the crossing of narrative storytelling and scientific research by media coverage of scientific findings.

3 The Way of Telling a Story: Issue Life-cycles

In general, several scientific disciplines and sub-disciplines are dealing with the way of telling a story. As we are trying to model the journalistic way, the main theoretical references in this chapter derive from journalism and communication research. These references are often modelling either the story telling process or the reconstruction of facts within a story. To combine them, knowledge about the interactions between process and content from other disciplines such as product life-cycles in economics and drama theory in humanities is used to complete the picture (Kolb, 2005). These approaches show that the analysis of processes has to take into account the changing of content-oriented variables and time as one variable. The way of analysis is longitudinal which basically means that data from different points in time can not be reduced to averages, sums or other indexes. The usual scientific method to reduce complexity in longitudinal data within the social sciences is to establish a model of phases each of which can be investigated entirely as one measurement point (Downs, 1972; Brosius, 1994 and 1998; Donsbach et al., 2005; Weßler, 1999).

In most of these studies, one important point is not referred to: There are not many scientifically developed and accepted methods of how to find different phases within a set of data. This is why studies vary widely concerning the way of distinguishing phases. Some researches take the entire data set and discriminate between as many phases as possible leading to a) enough data within each phase and b) enough phases to compare (Brosius, 1994). Others model only phases of the same length, which could by coincidence be adequate in some cases (Brosius, 1998). These studies, however, do not consider that time as a variable is nothing else but a symbol for changes within content. There is no plausible reason why media coverage of, for exmple, the lead problem should not be identical on every day of the week. The only reason for these differences lies in the development of the story and the events forming the story (Kolb, 2005).

The integration of the content of the phases to their distinction is not used very often, although it is rather fruitful: Weßler describes two different types of phases in his analysis of the media coverage of problems with drugs in Germany (Weßler, 1999). There are some phases referring to events like

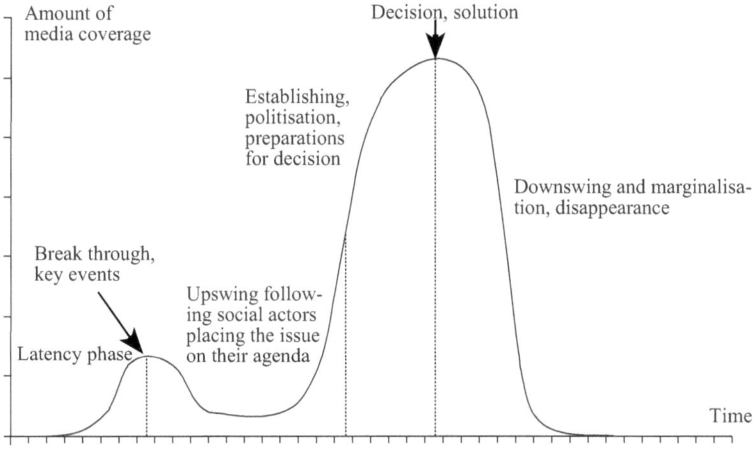

Fig. 1. Idealistic issue life-cycle (Kolb, 2005)

court decisions and others not related to any event. He found out that the 'event-phases' do usually produce more coverage.

Economics especially, but also some sub-disciplines of Communication Research tend to use life-cycle approaches to model the process of media coverage or public attention: For example, Downs has developed an issue attention cycle to explain the development of public interest in ecologic issues (Downs, 1972). Berens as well as Mathes/Pfetsch have modelled the media coverage of conflicts and crises (Berens, 2001; Mathes and Pfetsch, 1991). Dyllick has tried to establish an issue-cycle-model for marketing and public relations purposes (Dyllick, 1989). To integrate these approaches Kolb (2005) has put them on an identical basis: the diffusion of products and news respectively (Rogers, 1983).

The result is a phase model of issue life-cycles which tries to answer the following questions. Are there any recurring elements when the mass media bring up issues for discussion? How does the coverage of an issue change over time? How do the characteristics of the issues influence the process of media coverage? The theoretical framework integrates both, the dynamics and the content of media coverage. The cycle approaches are accompanied by well approved models of influences on media contents. Both are jointly used to elaborate categories for content analyses. The first empirical application of this model is presented in the next chapter.

In a first step the combination of the normal distribution and longitudinal empirical findings about media coverage of specific issues lead to an idealistic issue life-cycle. The distribution of coverage in time normally has more than one (local) extreme. Most of the coverage can usually be found around the point in time when the decision of the problem is developed. The decision often takes some time so that the amount of coverage increases in the second half. The idealistic graph for an issue life-cycle is shown in figure 1 (Kolb, 2005).

Derived from different phase models, Kolb worked out five phases:

1st The latency phase is characterised by a low level of coverage. A key event usually exemplifies the main problem at the end of this first phase. This event can be seen as a break through for the issue in question.

2nd During the upswing phase the issue becomes more and more important: the amount of coverage increases rapidly – at least after a first decrease directly after the key event.

3rd The increase of coverage loses speed during the establishing phase, probably due to a ceiling effect. The decision or solution of the problem is prepared and presented at the end of this third phase.

4th The downswing phase shows a significant decrease of coverage directly after the decision has been published. Whenever the implementation of the problem's solution happens after the decision, this fourth phase ends with the coverage of the implementation process.

5th The decrease leads directly to the marginalisation phase. At this time, the issue disappears from the media.

Besides individual, functional, organisational, structural, and other journalism system-immanent factors having an influence on the content of media coverage, Shoemaker/Reese identify external factors affecting the media story (Shoemaker and Reese, 1996). Big players in this context might be professional public relations as well as science. To ascertain all the possibly influencing factors the empirical analysis takes into account findings from four areas of Communication Research: Agenda-setting spots on different actors having an impact on the media's agenda (McCombs et al., 1997; Dearing and Rogers, 1996). The concept of newsworthiness emphasises the importance of the events being covered. Furthermore, it focuses on relevance, surprising facts, damage, negativity, and many more aspects (Galtung and Ruge, 1965; Staab, 1990). Risk communication analyses have identified conflict lines within journalistic argumentation and politisation as central factors influencing the coverage of an issue (Kafka and von Avensleben, 1998; Neverla, 2003). The framing approach models media frames changing over time. This might also focus on argumentation within the coverage (Scheufele, 1999, 2000 and 2003).

This social science approach to the story telling of mass media can be completed by regarding the humanities. Burkhardt develops a model for media scandals[4] derived from drama theory and discourse analysis (Burkhardt, 2006): The macrostructure of journalistic narration is a complex composition

[4] Details concerning the empirical case study are published in Burkhardt's "Der Medienskandal", cf. (Burkhardt, 2006). The basis for this case study is about 300 articles in German newspapers and magazines from the year 2003 and scandalizing the Vice President of the Central Council of Jews in Germany and TV host Dr. Michel Friedman.

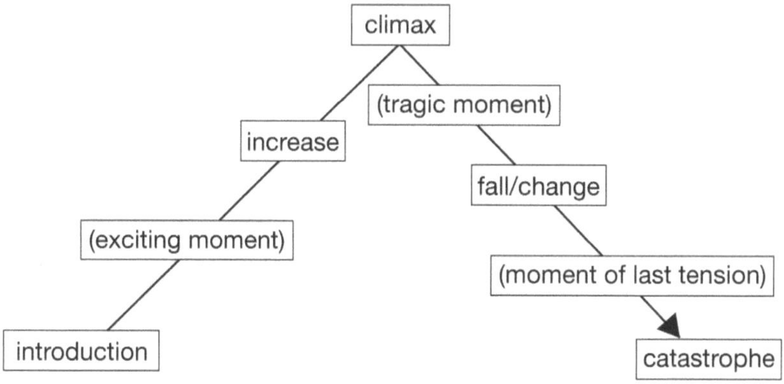

Fig. 2. Pyramidales phase principle after Freytag following Aristotle (Freytag, 1965)

of several storylines which build particular sequences. The media coverage passes through specific phases which are put on the public agenda much like a drama on stage in several acts. It follows the pyramidal structure principle of initial situation, development and dissolution which Aristotle designed in his poetics' 350 BC for the classical tragedy.

Aristotle's concept is the ideal of a closed drama which begins without preconditions, finishes with a definite end and builds one unity of story (Pfister, 2001/1997). Unity in this context means that there is one central storyline which starts in the first phase and ends in the last phase. This dominant episode has several sub-sequences, but only the main storyline has to follow the concept of closed drama in the tradition of Aristotle. For this main action Aristotle demands a clearly exposed starting situation which is based on a visible situation of circumstances. From these a conflict between transparently formed antagonistic persons develops and leads to a clear, action locking solution.

Gustav Freytag has brought the pyramidal design principle onto a simple succession formula from introduction, (optional) exciting moment, increase, climax, tragic moment, fall/change, (optional) moment of the last tension and catastrophe (cf. Fig. 2) (Freytag, 1965/1863).

The functions of the introduction phase in drama theory correspond with Kolb's empirical findings for the latency phase (Kolb, 2005). In this phase the setting of location, time and living conditions of the hero takes place and a first introduction of the topic and its context is given. At the end of the introduction phase, there is the option that an exciting moment starts. This exciting moment can be a crucial action of the protagonists or antagonists and can have the function of a key event. The action is set in motion, the main persons have shown their being, and the participation is lively. With a given direction, mood, passion and complexity lift itself. Nearby new action sequences are opened. The opponents or fellow combatants of the hero, who are not yet introduced, now get a mediated action area and the opportunity to interfere in the happening (Burkhardt, 2006).

The increase phase (of the drama) or upswing phase (of the journalistic storytelling) culminates finally in a climax or high point of the drama the models are also congruent here. As Freytag describes, the climax depicts the solution of the fight, which has ascended through the first part of the drama (Freytag, 1965). Therefore, at this time all action sequences usually reach their top. Thereafter, the action drops. Freytag states that the downswing phase is the most "difficult" part of the drama (Freytag, 1965): Until the climax is reached, the participation was bound to the direction of the main characters. After the act a break develops and the tension must be excited anew, hence new forces have to be activated. Perhaps even new roles have to be introduced in order to get the audience involved again. Thus there is the danger that the recipients lose their interest after the climax and it requires a new moment of suspense or several suspense-creating moments, in order to hold or win their attention again. Kolb's observation that in this phase the reports are made by the "implementation of the solution" corresponds with Freytag's idea, that the function of this phase lies in the illustration of the main concept which refers to action (Kolb, 2005; Burkhardt, 2006): "The core of the whole one, of the idea and guidance of the action step out powerfully, the spectator understands the connection of the occurrences, sees the last intention of the poet, he should addict itself to the highest effects and in the midst of his participation start to measure the artwork with his knowledge, with his cosy inclinations and needs" (Freytag, 1965).

The following "catastrophe" of media coverage is the finishing plot which is called exodus in poetry. It does not inevitably mean a disaster in the usual modern sense of the word. The conclusions of the journalistic narration thereby have the task of showing the meaning of the story to the audience. Moreover, they have to remind the recipients that nothing coincidental or unique is represented, but rather something poetic and of general importance.

The similarities between the ways of telling stories on stage (more fictional) and in media coverage (more factual) indicate that journalism is not only a system of researching, selecting and distributing themes from the social system and its environment. Journalism can be seen as a producer of popular culture, which means journalism works as "narrative generator of a common cultural understanding" (Klaus and Lünenborg, 2002; Dahlgren, 1986; Renger, 2000). Media coverage interacts in a 'struggle over meaning' between the narration and the social identities (Luthar, 1997). Therefore, even the media coverage of scientific findings can be read as meta-stories of reality. This kind of media coverage must not only be analysed as 'objective' coverage but also as a phenomenon of popular culture.

The different concepts of issue life-cycles show that the staging of scientific findings on the public media scene produces structures of meaning for the audience. They are part of the cultural system that wattles identity (Hall and duGay, 1997). They transform society to community by transmitting self-images and telling stories about common scientific problems and attainments that touch on the self-concept of common life.

4 The Way of Telling the Lead Story

In this part of the paper, three short versions of the media coverage of the lead problem are summarised, reconstructed, and interpreted. Despite the similarities in the dynamics of coverage following the idealistic model described in part three, the three national journalistic stories analysed in Germany, France, and the United Kingdom have their own specificities[5].

The German issue life-cycle starts back in the 1960s. In 1965, the palaeontologist Patterson presents his research results proving that the lead concentration in human blood is anthropogenic (Patterson, 1965). The use of lead as a resource for the development of anti-knock gasoline, colours, etc. has caused the increase of lead concentration in the atmosphere and in humans and following the critics, to a hazardous level. These findings are covered by some of the German media, especially the "Frankfurter Allgemeine Zeitung". During the latency phase, the scientific findings and the possible dangers are the focus of the German press.

The German socio-democratic and liberal government addresses the subject virtually supported by the debate about emission reduction in California. The first and second law on the reduction of lead in petrol are discussed respectively in 1971 and 1975. These decisions can be seen as the key events of the issue life-cycle marking the issue's breakthrough. The amount of media coverage reaches a first high. The content concentrates on the political debates and decisions as well as on the industry demanding tax reductions for clean cars and subventions for their investment necessary to produce clean engines.

The upswing phase is characterised by a change of the main frame within the coverage. Whereas the possible danger to public health followed by political debates were previously in the centre of interest, the ecologic problem of acid rain and the 'Waldsterben' becomes more and more central from 1978 to 1981. Technically speaking, the focus of TEL concentration reduction transformed into a debate about lead free petrol and the introduction of catalytic converters for clean car emissions. Once again, politicians added this topic to their agenda, which can be exemplified by the rise of the green party in Germany. The car industry would also benefit from the introduction because the expensive double development of models for a global market would become obsolete taking into account that Japan and the United States had already

[5] Details concerning the empirical case study are published in Kolb's "Mediale Thematisierung in Zyklen", cf. Kolb (2005). For reasons of readability, this work is not cited anymore during this chapter anymore. The study is based on about 1400 articles in German, French, and British newspapers and magazines from 1965 until 2000.

introduced catalytic converters and lead-free petrol. In addition, the tax reductions on clean cars pushed up the sales of new cars (VDA, 1971, 1991, 2001 and 2004). This time, however, it was the mineral oil producing industry which was among the critics, and demanded subventions to compensate for the investment in new gasoline pumps and tanks.

The political decision of the German government and the EC/EU-agreement to introduce lead-free petrol and catalytic converters marks the resolution of the problem at the end of the establishing phase. Further problems concerning car emissions such as the sulphur problem or the current discussion about soot derived from diesel engines are covered far less in comparison to this issue life-cycle. Furthermore, the subsequent EU regulations (e.g. the 1989 regulation) lead to some media attention in Germany. However, the extreme coverage observed in 1983 is not reached again. The life-cycle has ended with a marginalisation phase in the 1990s evaluating the political solutions of the lead problem as an ecological victory.

In the French media, the issue is discussed differently. The starting point and content of the latency phase is the German debate on 'Waldsterben' and the first EC/EU-regulations. Thus, the issue is politically framed in the first articles on the lead problem. A second characteristic of the French journalistic story is the criticism of the German political actions in Germany as well as in the EC/EU. French journalists tend to interpret the ecological movement as campaigning of the car industry. They even undertake journeys to the German forests in order to be able to call the Germans' fear of acid rain and 'Waldsterben' a myth based on false empirical evidence. The key event for the probably unfinished French issue life-cycle is the 1989 EU-regulation introducing the US-standards for car emissions in Europe.

At the end of the upswing phase, the French government presents its own Air quality report in Paris in 1996. Supported by initial scientific findings, the French begin to realise ecological problems especially in the big cities. Even today, Paris, Lyon, Marseille, and other places in France suffer from dangerous concentrations of different poisonous substances. During the establishing phase, the French media begin to give more attention to ecological experts presenting their studies. In brief, the evolution of the story is quite opposite to the German issue life-cycle: A political and economic issue becomes a scientific issue facing health and environmental problems until the year 2000. In this year, the French government presents its environmental plan. Additionally, the empirical data set ends in 2000 so that the interpretation of this point in time as the climax could be overcome by further research including the time after 2000.

In British media coverage, the framing is relatively constant over time. The main aspects of the debate as well as the mainly cited actors come from the political system and the industries involved. Particularly during the latency phase and the marginalisation phase, the British press focuses on the health hazard caused by car emissions. The environmental problems are not discussed at all, which might be due to the fact that the United Kingdom is an island state: Winds from the West are blowing most emissions in the direction of Scandinavia and Central Europe so that people in big cities suffer from huge amounts of emissions whereas the forests are not too exposed to heavy emissions. The key event at the end of the latency phase is the ninth report of the *Royal Commission on Environmental Pollution* which focussed on the lead problem. After its presentation in June 1983, the upswing phase is dominated by a journalistic service for car drivers who are afraid that their old cars may be damaged by the new lead-reduced petrol. In November 1988, the lead-free week and the unleaded petrol bill are the main topics in the British press. In March 1989, the establishing phase leads to a climax. The preparation of the EU-regulation and a further tax reduction in the UK represent the political solution of the lead problem for the British press.

The upswing phase up to the end of 1990 and the following marginalisation phase can be characterised by some articles concerning possible health risks due to the additives replacing the TEL. Another short media attention phase in the 1990s is caused by the introduction of Swedish 4-star petrol, which, according to the critics, would destroy the engines of 40 per cent of the British cars. The special British framing consists of constant political and economic coverage and a huge volume of service-oriented articles explaining how people should deal with the changes decided by the government or the EU. Generally speaking, the presentation of scientific findings to add objectivity to a journalistic story can happen at any point in time. The German issue life-cycle can be seen as an example of a 'scientific framing', which initiates the story with research results and ends it with evaluation results. In France, the science input gains media attention in the middle of the process, whereas in the UK scientific findings are presented from time to time during the story telling process. Regarding the content, the newsworthiness concept shows that a journalistic story has to be surprising and often negative in its consequences. This can be found in the coverage of the lead problem: the media tend to tell a dramatic story.

5 Conclusion

This paper deals with the problems arising when journalists try to cover scientific findings. Although the main functions of journalism and science appear, at a first glance, rather similar, the way in which knowledge is gained as well as the way the results are published differs substantially between the two systems. Whereas objectivity in science is established by transparency as intersubjective understanding and reproducibility of the studies, journalism constructs objectivity by using specific strategies to pretend authenticity and truth.

The main point of the theoretical chapter leads to the interpretation that science deconstructs truth – or whatever is postulated as truth in order to measure and model it. In contrast, journalism constructs truth by the presentation of opposing points of view and possibilities of reality, by the presentation of facts that support these statements and by the simulation of objectivity through selecting and marking quotations. However, the separation of facts and opinion in media coverage can always only be pseudo-objective.

In the empirical chapters, it is emphasized that journalistic publications concerning one issue follow a life-cycle. This life-cycle is rather similar to the classic drama set. Therefore the media coverage of scientific findings has to be analysed as a phenomenon of popular culture producing structures of meaning for the audience. The usual context for a journalistic general interest publication is society, which means that scientific findings are presented as social problems. When scientists are capable of describing their findings as social problems they can generate media attention. Unfortunately, this can lead to a lack of further investigation by the journalists who tend to present only one scientific point of view. Contradicting results are often only published by scientific journals and ignored by journalists due to a missing social contextualization.

In summary, the task for scientists communicating their findings to journalists is to work out the social relevance of their studies whereas the journalistic task would be to search for other results and the scientific state of the art.

6 References

Aarebrot F, Bakka PH (1997) Die vergleichende Methode in der Politikwissenschaft. In: Berg-Schlosser D, Müller-Rommel F (eds.) Vergleichende Politikwissenschaft. Leske, Opladen:49-66

Berens H (2001) Prozesse der Thematisierung in publizistischen Konflikten. Ereignismanagement, Medienresonanz und Mobilisierung der Öffentlichkeit am Beispiel von Castor und Brent Spar. Westdeutscher Verlag, Wiesbaden

Brosius HB (1994) Integrations- oder Einheitsfach? In: Publizistik 39:73-90

Brosius HB (1998) Publizistik- und Kommunikationswissenschaft im Profil. In: Rundfunk und Fernsehen 46:333-347

Burkhardt S (2005) Boulevard-Journalismus. In: Weischenberg S, Kleinsteuber HJ, Pörksen B (eds.) Handbuch Journalismus und Medien. UVK, Konstanz:31-35

Burkhardt S (2006) Der Medienskandal: Theorie und Empirie aus diskursanalytischer Perspektive. Diss., Universität Hamburg. Cf. chapter 6

Dahlgren P (1986) Beyond Information: TV News as a Cultural Discourse. In: Communications, 2/1986:125-237

Dearing JW, Rogers EM (1996) Communication Concepts 6: Agenda-Setting. Sage, London

Donsbach W, Laub T, Haas A, Brosius HB (2005) Anpassungsprozesse in der Kommunikationswissenschaft. Themen und Herkunft der Forschung in den Fachzeitschriften Publizistik" und Medien & Kommunikationswissenschaft". In: Medien & Kommunikationswissenschaft 53:46-72

Downs A (1972) Up and down with Ecology. The "Issue-Attention Cycle". In: The Public Interest 28:38-50

Dyllick T (1989) Management der Umweltbeziehungen. Öffentliche Auseinandersetzungen als Herausforderungen. Gabler, Wiesbaden

Foucault M (1974) Die Ordnung der Dinge. Eine Archäologie der Humanwissenschaften. Suhrkamp, Frankfurt/M.

Foucault, M (1978) Dispositive der Macht. Merve, Berlin

Freytag G (1965/1863) Die Technik des Drama. Darmstadt

Galtung J, Ruge, MH (1965) The Structure of Foreign News: The Presentation of the Congo, Cuba and Cyprus Crises in Four Foreign Newspapers. In: Journal of Peace Research 2:64-91

Habermas J (1990/1962) Strukturwandel der Öffentlichkeit: Untersuchungen zu einer Kategorie der bürgerlichen Gesellschaft (Mit einem Vorwort zur Neuauflage). Suhrkamp, Frankfurt/M.

Hall S, du Gay P (1997) (eds.) Questions of Cultural Identity. Sage, London

Kafka C, von Avensleben R (1998) Consumer Perceptions of Food-Related Hazards and the Problem of Risk Communication. In: AIR-CAT 4th Plenary Meeting: Health, Ecological and Safety Aspects in Food Choice, 1, 1998: 21-40

Klaus E, Lünenborg M (2002) Journalismus: Fakten, die unterhalten Fiktionen, die Wirklichkeit schaffen. In: Neverla I, Grittmann E, Pater M (eds.): Grundlagentexte zur Journalistik. UVK, Konstanz: 100-113

Kolb S (2004) Voraussetzungen für und gewinnbringende Anwendung von quasiexperimentellen Ansätzen in der kulturvergleichenden Kommunikationsforschung. In: Wirth W, Lauf E, Fahr A (eds.) Forschungslogik und design in der Kommunikationswissenschaft. Volume 1: Einführung und Aspekte der Methodenlogik aus kommunikationswissenschaftlicher Perspektive. Halem, Köln:157-178

Kolb S (2005) Mediale Thematisierung in Zyklen. Theoretischer Entwurf und empirische Anwendung. Halem, Köln

Krotz F (2005) Neue Theorien entwickeln. Eine Einführung in die Grounded Theory, die Heuristische Sozialforschung und die Ethnographie anhand von Beispielen aus der Kommunikatorforschung. Halem, Köln

Laclau E, Mouffe C (1985) Hegemony and Socialist Strategy. Towards a Radical Democratic Politics. Verso, London

Luhmann N (1990) Die Wissenschaft der Gesellschaft. Suhrkamp, Frankfurt/M.

Luhmann N (2004) Die Realität der Massenmedien. VS Verlag für Sozialwissenschaften, Wiesbaden

Luthar B (1997) Exploring moral Fundamentalism in Tabloid Journalism. In: Javnost The Public, Journal of the European Institute for Communication and Culture, 1/1997:49-64

MagDougall C, Reid RD (1987) Interpretative Reporting. Macmillan, New York

Mathes R, Pfetsch B (1991) The Role of the Alternative Press in the Agenda-Building Process: Spill-over Effects and Media Opinion Leadership. In: European Journal of Communication 6:33-62

McCombs ME, Shaw DL, Weaver DH (1997) Communication and Democracy: Exploring the Intellectual Frontiers in Agenda-Setting Research. Mahwah, London:85-96

Meyer P (1991) The New Precision Journalism. Indiana University Press, Bloomington, Indianapolis

Rorty, RM (1967) (ed.) The Linguistic Turn. Essays in Philosophical Method. University of Chicago Press, Chicago

Mill JS (1843/1959) A System of Logic. London:648 ff.

Mott FL (1959) American Journalism: A History of Newspapers in the United States Through 260 Years: 1690 to 1950. Macmillan, New York

Neverla I (2003) "Aktuell ist ein Wald erst, wenn er stirbt." Der Beitrag des Journalismus zur öffentlichen Risiko-Kommunikation. 2003. In: http://www. journalistik.uni-hamburg.de/institut/publikationennever-la.html, download from 11/27/2003

Patterson C (1965) Contaminated and Natural Lead Environments of Man. In: Archives of Environmental Health 11:344-359

Pfister M (2001/1997) Das Drama: Theorie und Analyse. UTB, Stuttgart

Przeworski A, Teune H (1970) The Logic of Comparative Social Inquiry. Wiley, Malabar

Renger R. (2000) Populärer Journalismus. Nachrichten zwischen Fakten und Fiktion. Beiträge zur Medien- und Kommunikationsgesellschaft 7. Studien Verlag, Innsbruck Wien München

Rogers EM (1983) Diffusion of Innovations. Free Press, New York London

Rühl M (1980) Journalismus und Gesellschaft. Bestandsaufnahme und Theorieentwurf. v. Hase & Köhler, Mainz

Sandbothe M (2000) Die pragmatische Wende des linguistic turn. In: Sandbothe M (ed.) Die Renaissance des Pragmatismus: Aktuelle Verflechtungen zwischen analytischer und kontinentaler Philosophie. Velbrück, Weilerswist

Scheufele DA (1999) Framing as a Theory of Media Effects. In: Journal of Communication 49:103-122

Scheufele DA (2000) Agenda-setting, Priming, and Framing Revisited: Another Look at Cognitive Effects of Political Communication. In: Mass Communication & Society 3:297-316

Scheufele B (2003) Frames Framing Framing-Effekte. Theoretische Grundlegung, methodische Umsetzung sowie empirische Befunde zur Nachrichtenproduktion. Westdeutscher Verlag, Opladen Wiesbaden

Schmidt J (2003) Geschichten & Diskurse: Abschied vom Konstruktivismus. Rowohlt, Reinbek/Hamburg

Shoemaker PJ, Reese S D (1996) Mediating the Message: Theories of Influences on Mass Media Content. Allyn & Bacon, White Plains

Staab JF (1990) Nachrichtenwert-Theorie: Formale Struktur und empirischer Gehalt. Freiburg/Breisgau München

Tuchman G (1972) Objectivity as Stratetegic Ritual: An Examination of Newmen's Notions of Objectivity. In: American Journal of Scociology, Vol. 77:660-679

Tuchman G (1978) Making News. A Study in the Construction of Reality. Free Press, New York

van Deth JW (1998) Equivalence in Comparative Research, In: van Deth JW (ed.): Comparative Politics: The Problem of Equivalence. Routledge, New York:1-19

VDA (1971) Tatsachen und Zahlen aus der Kraftverkehrswirtschaft. Vol. 35. Frankfurt/M.:206-207

VDA (1991) Tatsachen und Zahlen aus der Kraftverkehrswirtschaft. Vol. 55. Frankfurt/M.:272-273

VDA (2001) Tatsachen und Zahlen aus der Kraftverkehrswirtschaft. Vol. 65. Frankfurt/M.:220-221

VDA (2004) Neuzulassungen. In: http://www.vda.de/de/aktuell/statistik/jahreszahlen /neuzulassungen/index.html, download from 19/02/2004

Weber R (1974) (ed.) The Reporter as Artist: A Look at the New Journalism Controversy. Hastings, New York

Weischenberg S (1998) Journalistik. Medienkommunikation: Theorie und Praxis 1: Mediensysteme, Medienethik, Medieninstitutionen. Westdeutscher Verlag, Opladen:51 ff.

Weischenberg S, Löffelholz M, Scholl A (1996) Journalismus in Deutschland. Westdeutscher Verlag, Opladen

Weßler H (1999) Öffentlichkeit als Prozess. Deutungsstrukturen und Deutungswandel in der deutschen Drogenberichterstattung. Westdeutscher Verlag, Opladen Wiesbaden

Wirth W, Kolb S (2004) Designs and Methods of Comparative Political Communication Research. In: Esser F, Pfetsch B (eds): Comparing Political Communication: Theories, Cases, and Challenges. Cambridge University Press, New York: 87-114, cf. p.:88

Wolfe T (1975) The New Journalism. Pan Macmillan, London

Fishing for Sustainable Policies – Political Causes of Over-fishing and Solutions for Sustainable Fisheries

Christian von Dorrien

Federal Research Centre for Fisheries, Institute for Baltic Sea Fisheries (IOR), Alter Hafen Süd 2, 18069 Rostock, Germany

1 State of World Fisheries

The intergovernmental Food and Agriculture Organisation of the United Nations (FAO) has one department concentrating on fisheries and fisheries-related questions. The FAO collects data on fisheries landings from countries worldwide in order to produce global landing statistics. Every two years, the FAO publishes an extensive report on the "State of World Fisheries and Aquaculture" (SOFIA). The most recent version was published in 2004 (FAO, 2004). As it normally takes up to two years to collect, process, analyse and publish the landing statistics, the most recent report from 2004 reflect the state of world fisheries in the year 2002. The SOFIA report can be found on the Internet under http://www.fao.org/sof/sofia/index_en.htm. Some of the main findings of this report are outlined below. All data on weight represent the live weight of fish.

Looking at the graph showing the world capture fisheries production, there is almost a five-fold increase in landings from the 1850's up to the year 2002. Although there has been a steep increase in production up until the 1970's, the landings do levels off in the 1990's. Recently, Watson and Pauly (2001) specifically analyzed the Chinese landing statistics. They found that over the last 10 years China had reported much higher landings than it actually achieved. The authors give different reasons for this over-reporting. If the world capture fisheries production is looked at without including the over-reported Chinese landings, it becomes obvious that from the mid-nineties onwards there has been a slight decrease in the world's capture fisheries production. Furthermore, if one takes into account that the fishing effort has increased through the use of improved technologies and as well as extending fisheries to once un-fished areas and stocks, such as for example deep sea fish stocks, this is an alarming signal. The situation is further exacerbated by the fact that nearly 30% of all marine fished stocks are either overexploited, already depleted, or recovering. More than 50% of stocks are fully exploited, which means that careful management is needed so as not to over-fish them. In fact only 25% of marine stocks are underexploited or moderately exploited. Only these stocks are perhaps able to satisfy the still growing demand for marine fish. Thirty years ago, nearly 40% of the stocks were in this good condition. Since then the proportion of the stocks being over-fished has drastically increased, however, over the last 8 years this development appears to be slowing down. In general, many fish stocks worldwide are not used in a sustainable manner and are in need of better fisheries management. In many cases this means a reduction in the fishing pressure as a first measure.

1.1 The Top Ten Species and Fishing Nations

From a total of nearly 85 million tonnes of marine landings, one species alone, the Peruvian anchoveta (*Engraulis ringens*), accounted for nearly 10 million tonnes and is by far the most important species. The second most important species, the Alaska pollock (*Theragra chalcogramma*), accounted for 2.7 million tonnes of the overall landings, followed by skipjack tuna (*Katsuwonus pelamis*) and capelin (*Mallotus villosus*) with 2 million tonnes each. As in previous years, China was the single most important fishing nation with a total production of marine and inland capture fisheries of 16.6 million tonnes. Peru, fishing mainly for anchoveta for the production of fish meal and fish oil, was with 8.8 million tonnes the second most important producer, followed by the United States with 4.9 million tonnes.

1.2 Developing Countries and Trade

A large share of the fish landings is traded internationally. The value of net exports of fish and fish products in developing countries has increased from less than 5 billions US$ in 1982 to more than 17 billion US$ in 2002. This is more than has been earned by the same countries for coffee, cocoa, bananas and other agricultural commodities put together (FAO, 2004).

To highlight the importance of fisheries and the trade in fish products for developing countries, it is a riddle to ask for the most important landing place for fresh fish in Germany. This is not, as one might expect, a harbour on the coast, but it is the Airport in Frankfurt, located right in the middle of the country. In comparison with the major fishing harbour Cuxhaven on the North Sea coast, where only 6,500 tonnes have been landed in 2001, Frankfurt received 28,000 tonnes by plane from 30 different countries – many of whom being developing countries. However, the majority of the over one million tonnes of fish which are consumed and processed in Germany, are "landed" and imported by trucks and ships, and mostly as deep frozen produce.

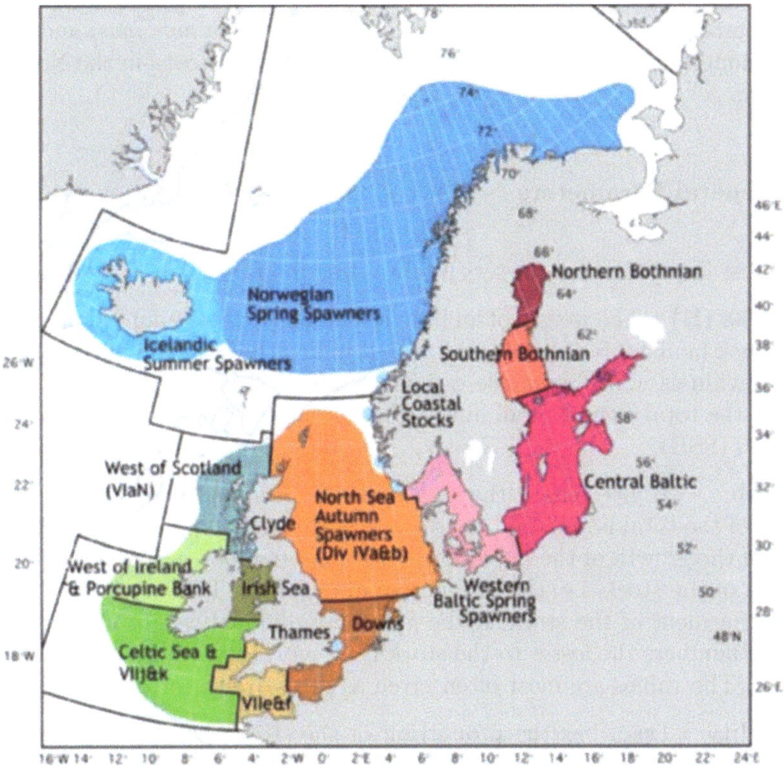

Fig. 1. Map showing the different stocks of the Atlantic herring (*Clupea harengus*). Graph: C. Zimmermann (Institute for Baltic Sea Fisheries, Rostock)

2 Short Introduction to Fisheries Management

2.1 What is a Stock?

When talking about the management of living marine resources, a lot of special phrases and technical expressions are used. Although it is not the intention here to replace the very good textbooks which are giving a good introduction into the wide field of fisheries management, there are some central terms for which it would be helpful to know at least a basic explanation.

A central and very often used word is that of "stock", and is used to describe a specific management unit. In most, but not all, cases it signifies a population of a given species which lives in a defined geographical area and shares similar biological parameters, such as growth, size at maturity, fecundity, and a similar fisheries mortality rate (see below). A thorough understanding of the biology of the population in question is needed in order to define these biological parameters. In many cases it is hard to measure these parameters directly and indirect estimations have to be used instead. Figure

1 gives an example of how the Atlantic herring (*Clupea harengus*) and its different populations are separated into several different stocks in the Northeast Atlantic.

2.2 General Parameters

The most important parameters used to describe the status of a stock are

Biomass (B) – The weight of an individual or the whole population or stock. The stock biomass is normally calculated indirectly from the mean weight of the individuals and abundance estimates. Most often the parameter is used to give the total weight of all adults in a population, or the Spawning Stock Biomass (SSB).

Growth – This gives either the gain in weight of an individual or the gain in weight of the total population or stock. A population can increase its biomass by both the growth of the individuals and – more importantly – by the reproduction of the stock, i.e. the production of offspring. The latter is also called the recruitment of the stock. If the gain by individual weight and reproduction outnumbers the losses to the stock from mortality, the stock shows a net growth. The values are most often given as rates over a certain time period.

Mortality – Describes the proportion of the stock that is lost due to the death of individuals, either by dying from old age or by predation by other animals. This proportion is called the natural mortality (M). The part of the total mortality rate (Z) caused by fishery is called fishing mortality rate (F), given over a certain time period. All the parameters Z, M, F are used in mathematical expressions to model the growth of the stock.

The fishing mortality rate (F) is the most important parameter when managing a fish stock. In nearly all cases, it is only the human activity of fishing which can be directly changed to "manage" the stock. The fishing mortality rate can range from 0 (no fishing) to high values of 1.0 or more per year. To reflect what is happening to the stock as precisely as possible, F should give the amount of all individuals in the stock which died as a consequence of fishing activities. These include not only those fish which are finally landed, but also those which passed through the meshes of the net and later died from their injuries and – most important – those individuals which are taken on board and are later thrown back to the sea (discarded), because they are not, for whatever reason, considered part of the landing. Such reasons could include, for example, that they are too small to be allowed for landing, or for other primarily marketing reasons. Therefore, it is clear that an underestimation of F is obtained if it is calculated only on the basis of the fish actually landed in the harbour. Furthermore, it is also important to distinguish F as the *rate* of fishing *mortality* in a model from the proportion of the fish stock which is caught, the latter being the *harvesting rate*.

2.3 Stock Assessment

The above parameters and many others are used in mathematical and statistical models to answer the following questions for the management of a fishery:

1. What is the current state of the stock?
2. What has happened to the stock in the past?
3. What will happen to the stock in the future under alternative management scenarios or options?

These resulting quantitative predictions are used to make scientific recommendations for catch ratios and to assist in the subsequent decisions taken by fisheries managers and politicians about the Total Allowable Catches (TACs) and quotas for the following year.

Basically, fishing is based on the fact that fish and many other marine living resources are producing much higher numbers of offspring (eggs and larvae) than are actually needed to sustain the population. This natural "strategy" already takes into account that most of the offspring will not survive long enough to replace the two parent individuals. In theory, a fishery should take only the excess production so as not to threaten the survival of the population as a whole. In fact, studies have also demonstrated that for many fish stocks the production rates even increase when the whole population is reduced to below its original, un-fished size. The reasons for this are numerous, and include possibe reduced competition and cannibalim or higher survival rates of juveniles. The problem, however, is that it is very difficult to find exactly the right balance. In many cases, over-fishing, combined with high natural variability and unpredictable bad conditions in certain years for the stock recruitment, has resulted in severe depletions of once highly productive stocks.

To adjust the management of many fisheries, a set of so called reference points has been defined. Historically, the Maximum Sustainable Yield (MSY) has often been misunderstood to mean maximize catches, and this unfortunately has led to the over-fishing of many stocks. To make fisheries more sustainable, i.e. to ensure that only the surplus production of a stock is used, a more careful approach is needed. In order not to threaten the possibility of a stock producing enough offspring, i.e. to ensure a sufficient recruitment potential, even in bad years, the so called Limit Reference Points have been defined. For example, the size of the spawning stock should not fall below the limit Biomass (B_{lim}), because at these low levels, the potential for sufficient recruitment is endangered. Corresponding to B_{lim}, a limit Fishing mortality rate (F_{lim}) is defined, that by no means should be exceeded so as not to risk the stock from falling below B_{lim}.

Very often it is difficult to directly measure the parameters required, and it is therefore necessary to use indirect estimations. Unlike in forestry, a fisheries biologist can, in most cases, not go out to sea and count all the individuals directly. In addition, living marine resources are highly influenced by the natural variability of environmental factors such as temperature, light, and food supplies. Therefore, as a consequence, even with the most sophisticated mathematical models and most precise measurements and estimations very often have a high proportion of uncertainty remaining when in comes to predicting future catches in the following year. To account for this uncertainty and to minimize the risk of exceeding the limit reference points, so called precautionary reference points have been defined (B_{pa}, F_{pa}). If these are crossed, then measures to reduce the forthcoming fishery should already be taken on this early stage. Depending on the existing knowledge and the predictability and risk level involved, the reference points are lying farther from or closer to the limit reference points.

2.4 ICES-Advice & Reference Points

The scientists working in ICES have defined different categories of a stock, depending on how the actual spawning stock biomass and fishing mortality rates are in relation to the precautionary and limit reference points. For example, if a stock's spawning stock size is above B_{pa} and the fishing mortality is below F_{pa}, the stock has its "full reproductive capacity" and is "harvested sustainably". However, if the fishing mortality is higher than F_{lim}, the stock is classified as "harvested unsustainably". If the spawning stock size is below B_{lim}, the stock is classified as "suffering reduced reproductive capacity". With this classification scheme it is possible to define and visualize the status of a stock, depending on its actual size and fishing pressure.

3 Fisheries in the European Union (EU) – Status and Problems

3.1 The EU Common Fisheries Policy

All member states in the European Union (EU) have handed over their responsibility for marine fisheries to the EU. Consequently, the fishing sector is a EU responsibility. The main body organizing the day-to-day business of fisheries management is the EU Fisheries Commission. The main decision body is the EU Council of Fisheries Ministers. To regulate marine fisheries and EU fishing vessels in European waters the "Common Fisheries Policy" (CFP) was adopted in 1983 for the management of fisheries and aquaculture. As many problems have been detected over the years, two reviews took place in 1992 and 2002. A revised CFP is now in force since 2003.

The CFP has four main Components:

- Conservation & Enforcement Policy: The main policy to manage fishing by EU vessels in EU waters.

- Structural Policy: In this policy, the size and structure of the EU fishing fleets are regulated as well as the distribution of relevant subsidies.
- Marketing Policy: To ensure that the fish processing industry and internal markets are sufficiently supplied.
- External Fisheries Policy: To establish contracts with non-EU countries (so called third countries), and especially with developing countries in order to establish access for EU vessels to their waters.

Status of EU fisheries in relation to World Fisheries

Although the EU member countries with the largest catches – Spain and Denmark – are not among the top-ten producers, the catches of all EU member states taken together have been ranked third on a global scale in 2001.

The International Council for the Exploration of the Sea – ICES

ICES is not a regional fisheries organisation like the NEAFC (Northeast Atlantic Fisheries Commission), and it has no regulatory power. Its main function is in formulating scientific advice which can then be used by the EU to set up quotas for waters in the Northeast Atlantic. ICES was founded in 1902 when it became quite apparent that fish stocks are limited and the first signals of over-fishing where detected. ICES coordinates and promotes marine and fisheries research in the North Atlantic in order to gain knowledge which will ultimately be used in producing advice. It is an intergovernmental organisation with 19 member states, including all relevant fishing EU member countries, as well as Iceland, Norway and Russia.

3.2 The Annual EU Quota Circle

Each year in December, the EU Council of Fisheries Ministers meets to agree and decide upon the Total Allowable Catches (TACs) for the following year. The whole process could be called and described as the annual EU Quota Circle.

The starting point in this circle is the scientific recommendation made by the ICES working groups. The scientists in the ICES working groups use data and other findings to assess the status of the stocks and to produce the scientific advice for future catches. One part of these underlying data is gathered by independent scientific surveys carried out on research vessels, where normalized catches of adult, juvenile and larval fish as well as hydro-acoustic surveys are undertaken to measure the actual stock sizes as directly as possible. However, as the resources for scientific investigations and research vessels are rather limited, the majority of the data, information and analysis is derived from commercial landings. The resulting scientific advice is based on the concept of sustainable fisheries and recognizes the precautionary reference points for fishing mortality and spawning stock biomass, to avoid any over-fishing and to ensure that the stocks are keeping their potential to produce enough offspring.

The formulated advice is then used by the EU Fisheries Commission to produce a proposal for the total allowable catches (TACs) in the following year. In many cases, such proposals are higher than those advised by the scientists and are justified on economic grounds. Finally, the Comission's proposals have to be adopted by the EU Fisheries Council, and it is these fisheries ministers who have the political power to make a final decision. Nearly every year and for many of the quotas, there is a lot of political bargaining and all too often the agreed TACs are much higher than those originally recommended by the scientists. The main reason for this is that the politicians are trying to satisfy the wishes of the fishing industry for higher catches. Finally, the fishing fleets of each country are allocated their own national TAC. In many cases, and especially where the quotas for many stocks have to be reduced, these fishing fleets are catching more than their allocated TACs. One – nearly unavoidable – reason for this is the by-catch of individuals – mainly juveniles which are smaller than the allowable size for landings. These have to be thrown back into the sea or "discarded" – right after the sorting of the catches onboard. The problem is, therefore, that these catches are neither counted in the quota nor do they have to be reported back to the scientists. Illegal catches are another problem, i.e. catches which are "black landed", and are not reported. Both these processes are resulting in a higher fishing mortality of the stocks and further lead to flawed and biased data sets which have to be used as the basis for the advice for the following year TAC's. As a consequence, the quota circle starts with an added uncertainty to the data sets.

3.3 EU CFP and the Status of Stock Conservation in EU Waters

As a result of the inconsequent quota policy, of the around one hundred fish stocks which are assessed annually by the ICES scientists, the spawning stock biomass of over 20 percent are outside safe biological limits, as has been demonstrated by Hammer & Zimmermann (2003) and shown in Fig. 2.

In addition, around 10 percent of the assessed stocks are harvested outside safe biologic limits. This picture didn't change between 1996 and 2002. Even more disturbing, the authors found that the average annual deviation of the official TACs from the scientific TACs proposed by ICES, increased from nearly 20% in 1987 to more than 30% in 2003, and in some years were lying as high as 50% (Fig. 3). Under these circumstances, it is no wonder that many of the EU fish stocks are over-fished. Whereas the term "outside safe biological limits" does not mean a threat of biological extinction of the affected stock, it does signify considerably lower catching potentials, or in other words, a waste of resources.

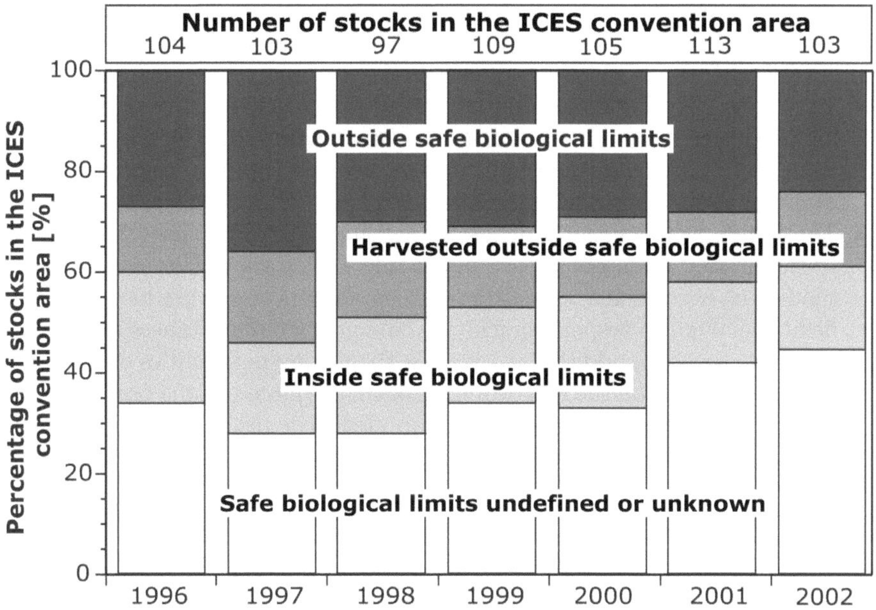

Fig. 2. Development of stocks in the ICES-Convention area, 1996–2002, in relation to safe biological limits. Graph from Hammer & Zimmermann, 2003 (with kind permission)

Why has the CFP not been able, in many cases, to ensure profitable fisheries on abundant fish stocks over the two decades of its existence? A main reason could be found in the conflicting aims which are formulated in the policy:

- On the one hand, fish stocks should be conserved, but at the same time fishing activities should be promoted. This is why catch quotas are, in many cases, set higher by the politicians than are scientifically recommended.
- Whereas the production (i.e. catch) should be modernized (i.e. enhanced), the fishing effort has to be limited. To adjust the fleet capacity to the available resources, and thereby reduce its size by scraping vessels, is conflicting with the aim of maintaining employment. With the help of many subsidies, the EU fleet has been built up to sizes which the experts have calculated as being 40 percent too large for the available fish resources.
- At an EU level, the implementation of conservation measures is called for, however, each member state is responsible for both monitoring and implementing sanctions, and many member states are falling short of doing this. The most recent examples are the high fines of 57 million Euro which France has to pay for every 6-month period from July 2005 onwards, until it fully complies with the obligations of the European Court of Justice. In 1991, the Court already ruled that France had failed to enforce

measures designed to protect undersized fish (below the legal minimum size), especially for hake (EU Commission Press release from 1. March 2006 http://europa.eu.int/comm/fisheries/news_corner/press/inf06_12_en.htm).

- There are also conflicts within the fishing sector itself. Although income for fishermen should be ensured, the processing industry relies to more than fifty percent on imported raw fish, because supplies from the EU are declining and imports are increasing.
- The policy asks for the acquisition of fishing rights in third world countries, without threatening the sustainable exploitation. However, many of the fishing agreements, especially with developing countries, have not been flexible enough to respond quickly to emergency circumstances such as decreasing stocks. In addition, very often the fishing possibilities offered were not always based on the real evolution of the resource and the precautionary approach is only rarely mentioned or applied. As a result, EU fishing vessels have at least been partially responsible for the overfishing of resources off the coasts of developing countries.

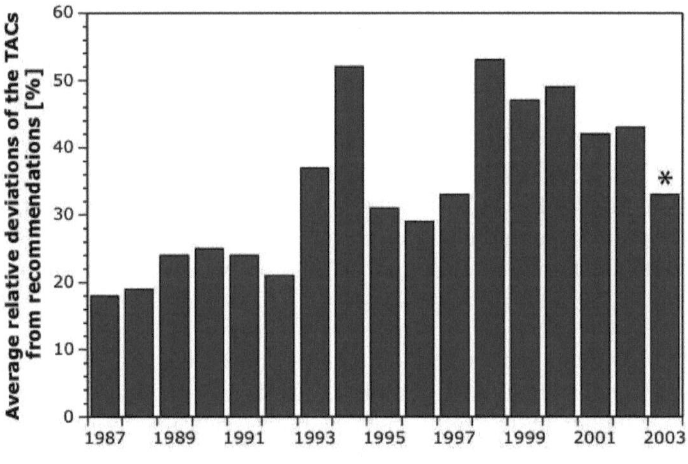

Fig. 3. Average annual deviation of the official TACs from the scientific TACs which were proposed by ICES (four extreme values between 1000 and 2200% have been omitted). *Provisional figure for 2003. Graph from Hammer & Zimmermann, 2003 (with kind permission)

3.4 Two Examples of the Consequences of the CFP

The History of Cod in the North Sea

To demonstrate that policy is unable to ensure stocks remain in a good condition, and at the same time to show that the basic instruments are in place to build up existing stocks back to high levels, can be illustrated using two examples.

The fate of cod in the North Sea demonstrates the decline of a fishery that for some thirty years ago was one of the most valuable. Whereas the decrease of the cod stock in the North Sea could at least partially be explained by changes in the environment, the catastrophic situation of the stock at the moment has also been caused by wrong policies and management.

In Fig. 4, the history of cod biomass and fishery in the North Sea for the years 1965 until 2001 is shown. It can easily be seen that in the sixties biomass was high and fishing mortality relatively low. Later on, fishing mortality steadily increased, whereas the size of the spawning stock rapidly decreased. In the most recent years, when the spawning stock size had decreased to levels of only a fifth the size in the sixties and well below the limit reference points, the fishing mortality was still above the limits. The cod stock in the North Sea is still at historic low levels, with no signs of recovery. Furthermore, and in spite of scientific recommendations to stop all fishery for cod in the North Sea, there are still considerable catch quotas set by the EU Fisheries Council.

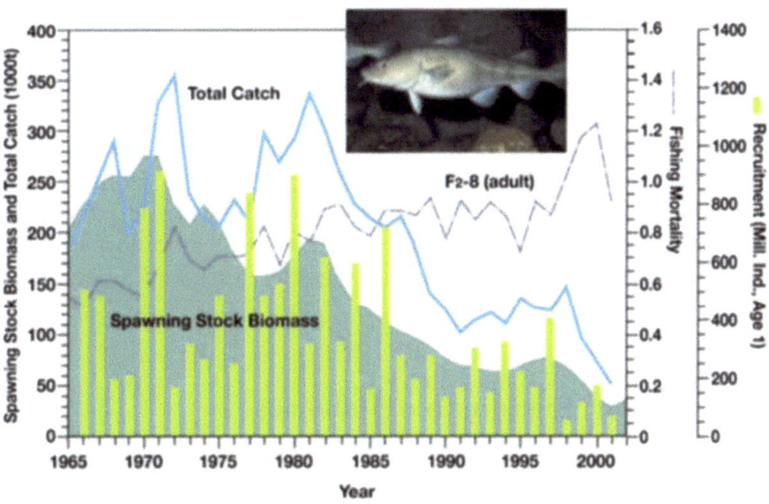

Fig. 4. The decline of the North Sea cod stock and the increase in fishing pressure. Graph: C. Zimmermann (Institute for Baltic Sea Fisheries, Rostock)

Herring: Lessons finally learned

Figure 5 shows the fate of herring stocks and its fishery in the North Sea. Initially, very high spawning stock biomasses occurred in the early sixties. However, when the fishery steadily increased, resulting in extremely high fishing mortality rates, the stock collapsed in the mid-seventies, resulting in the

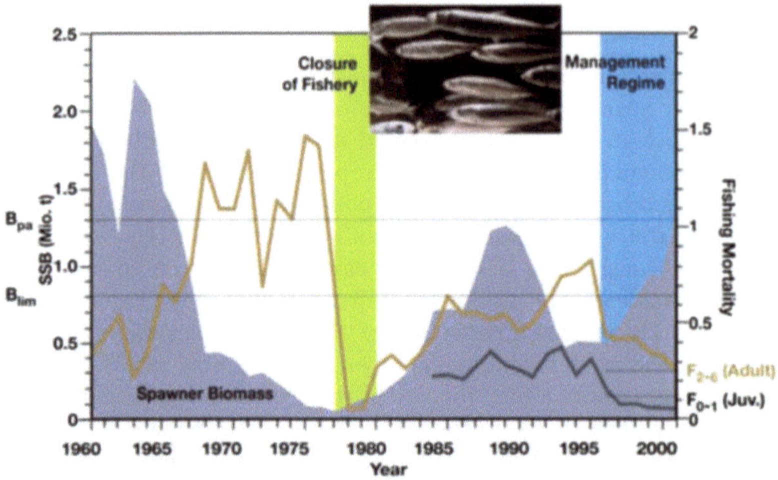

Fig. 5. Development of the herring stock in the North Sea. Graph: C. Zimmermann (Institute for Baltic Sea Fisheries, Rostock)

complete closure of the fishery. Luckily, because of strict enforcement of the catch moratorium and some good years of herring recruitment, the stock recovered quickly, so that the fishery could be re-opened in the early eighties. Unfortunately, however, a high fishing pressure put the herring stock again at risk. Therefore, in 1996, a tight management regime for the herring fishery in the North Sea was set up by the EU and Norway, leading to fishing mortalities below the precautionary limits. As a consequence, the spawning stock biomass increased over the precautionary limit reference sizes. Nowadays, the herring stock in the North Sea is considered as a healthy stock and an economically viable fishery.

4 The Solution: Sustainable Fisheries

In theory, the basic requirements for carrying out a fishery in a sustainable manner can be easily identified:

- No Over-fishing: The resources are used in such a way that the stock can sustain the fishery over an unlimited period of time and that the catches are stable on a high level. It is most important, however, that the ability of the stock to produce sufficient offspring – i.e. to show a high recruitment – is not affected.
- Minimized impact on Ecosystems: An economically viable fishery could only take place in a functioning and healthy marine environment. Therefore, the by-catch of unwanted species, especially of sea turtles, sea birds and sea mammals, but also other fish species that are sensitive to over-fishing,

such as sharks, should be reduced as much as possible. Fishing methods which are seriously affecting marine habitats, such as, for example, fishing with heavy bottom trawl gears over coral reefs, should be avoided and replaced by other less impacting fishing methods. Environmentally sensible and unique habitats, such as sea grass beds and deep sea corals, and other important areas such as spawning and nursery areas should be protected. In general, ecosystem aspects should be integrated into the assessment and management of a fishery.

- Effective Management: This ensures that the above mentioned principles are translated into concrete measures which are implemented and followed by the fishery. This covers also the participation of all stakeholders and a fair allocation of quotas. It includes measures for control and surveillance, as well as sound fisheries research to deliver the required data and knowledge.

Some basic requirements to ensure a sustainable fishery include:

- A thorough and sound assessment of the stocks to be used and their productivity.
- The fishing capacity (number and size of vessels) and the fishing effort used by these should be in balance with the available resources and their productivity.
- Selective gears/equipment and catching methods should be promoted and used.
- The existence of Marine Protected Areas is useful for protecting sensitive habitats and environmentally important areas and species. In many cases they might also function as an insurance against management failures and unforeseen changes, caused, for example, by climatic and environmental variability.

4.1 Implementing Sustainable Fisheries on the International and Regional Level

Based on the worldwide acceptance that the environment should be protected and used in a sustainable manner so as to ensure its existence for coming generations, relevant instruments have also been developed in the field of marine fisheries. On an international level, two main initiatives in the 1990's are to be mentioned here: The development of the voluntary "Code of Conduct for Responsible Fisheries" by the FAO and the adoption of the "Agreement on Straddling Fish Stocks and Highly Migratory Fish Stocks" by the United Nations. On a regional level, the reformed EU Common Fisheries Policy has been place since 2003, and has as one of its central objectives the implementation of sustainable fisheries.

4.2 The FAO Code of Conduct for Responsible Fisheries

In 1995, the Twenty-eighth Session of the FAO Conference agreed on the Code of Conduct for Responsible Fisheries (FAO, 1995). Fisheries and aquaculture provide an important source of food, employment, income and recreation, and millions of people depend upon fish for their livelihoods. Everyone involved in fishing must help to conserve and manage the world's fisheries, to ensure that there is going to be enough fish for future generations. The FAO Code of Conduct focuses on helping countries develop or improve their fisheries and aquaculture industries. It is important to note that the FAO Code of Conduct is voluntary and not mandatory. It is aiming at everyone working in, and involved with, fisheries and aquaculture. It has been translated into many languages, of which five are official. More detailed information about the FAO Code of Conduct and its implementation can be found at http://www.fao.org/fi/agreem/codecond/codecon.asp.

The role of the FAO is to technically support the activities of the governments who have the responsibility of implementing the FAO Code. In doing so, governments should incorporate the Code's principles and goals into their national fishery policies and legislation, encourage fishing communities and industry to develop *codes of good practice* which further support the goals and purpose of the Code of Conduct.

The main issues of the FAO Code of Conduct are responsible fisheries management; development of aquaculture, linking of fisheries with other coastal zone activities, and the processing and selling of the catch. The importance of countries to cooperate with each other in these issues must be stressed.

International Plans of Action (IPOA)

Within the framework of the FAO Code of Conduct and to deal with actual and special problems in fisheries, the FAO has developed within the framework of the FAO Code of Conduct the voluntary instruments of the so-called International Plans of Action (IPOA). To date, there are four in place. Two are aiming at the protection of seabirds (IPOA for Reducing Incidental Catch of Seabirds in Longline Fisheries) and sharks (IPOA for the Conservation and Management of Sharks). The third one involves an initiative to reduce the world wide overcapacity of fishing fleets (IPOA for the Management of Fishing Capacity). The fourth focuses on preventing, deterring and eliminating illegal, unreported and unregulated fishing (IPOA-IUU) (FAO 2001). It aims to stop a type of fishery which is sometimes called "pirate fishing". IUU fishing is of increasing concern as it undermines all efforts to achieve sustainable marine fisheries. To date, the respective international policies and instruments in place have been ineffective at stopping IUU fishing. A major part of this problem is that vessels are fishing under flags of convenience of states which do not control the fishing activities of these vessels.

A fishery is called "illegal", if, for example, vessels are fishing in foreign waters without permission, or otherwise breaking existing laws, regulations or international agreements. This is mainly the type of fishery which is known as "pirate fishing".

"Unreported" catches are those where a vessel is not reporting its landings, or where the place of the catches is misreported. These types of catches are often also called "black landings".

In an "unregulated" fishery there are no regulations, no licencing systems or any other type of controls in place. This could be the case for large areas in the open sea or for a fishery which uses stocks for which no quotas are in place.

4.3 The UN Fisheries Agreement

In 1982 the United Nations Convention on the Law of the Sea (UNCLOS) was adopted and entered into force in November 1994. It provides coastal states with, among others, the exclusive sovereign rights to explore, exploit, conserve and manage fisheries within 200 nautical miles (370.4 kilometres) of their shores. This zone is called the Exclusive Economic Zone (EEZ). By contrast, a state has full sovereignty in its own Territorial Waters which normally extends 12 nautical miles offshore. UNCLOS, however, contains gaps when it comes to the rights of states regarding highly migratory fish stocks and straddling fish stocks on the high seas. Straddling fish stocks are so named because they straddle or migrate between the outer limit of national fisheries waters of coastal States and the adjacent high seas (i.e. outside the 200 mile limit). Highly migratory fish stocks migrate through the high seas and through the EEZ of coastal States. Typical examples include several species of tuna. In many cases, these stocks are prone to over-fishing, and there are conflicts and disputes over the fisheries of these stocks and their management.

The UN Fisheries Agreement on Straddling Fish Stocks and Highly Migratory Fish Stocks (UNFA) was adopted in August 1995 by a UN conference. It brought major advances to the international fisheries law. Firstly, the flag states are responsible for the activities of their vessels. The UNFA also offers new instruments to enforce fisheries laws and regulations on the high seas. Finally, it has set up measures for the settlement of disputes between states on the management of a certain fishery.

Another major step was the implementation of the precautionary approach in the context of international fisheries management. This precautionary approach states that the absence of scientific information cannot be used as an excuse for failing to establish conservation and management measures. States and their fisheries managers are required to set stock-specific precautionary "reference points", as well as target levels for the fishing effort; they are also required to be "more cautious" when information is uncertain, and to ensure the conservation of other species.

It took more than six years before the 30th state has ratified the UNFA and it subsequently entered into force in December 2001. The EU only ratified the UNA in 2003. Today (2005), 59 nations are signatories to the UNFA, and 52 nations have ratified it. However, from the top ten nations of marine capture fisheries, only four have ratified the UNFA (United States, India, Russia and Norway). Another two (Indonesia and Japan) have signed but not ratified it – meaning, they are not obliged to abide by its rules. Four nations, among which are the top producers China and Peru as well as Chile and Thailand have not even signed the UNFA.

Aside from the missing signatures of top-fishing nations, there are still other issues which have to be resolved, including the need for an effective implementation of UNFA as well as major questions with regard to the control and surveillance on vast ocean areas.

Nevertheless, both the UNFA and the FAO Code of Conduct have been two important steps forward on the way towards more sustainable marine fisheries. They are also very helpful instruments for making national fisheries laws and management more sustainable.

4.4 The Reformed EU Common Fisheries Policy

The basic regulation of the EU Common Fisheries Policy (2371/2002) was reformed and finally adopted in 2002 and entered into force in 2003. The main objectives are formulated as:

- a sustainable exploitation of living aquatic resources,
- the application of the precautionary approach with the aim of protecting and conserving the resources and of minimising the impact on marine ecosystems,
- the implementation of an Ecosystem-based management, and
- efficient fishing activities within economically viable and competitive fisheries.

In order to achieve a conservation and sustainability of resources, several instruments are presently being planned, and these include, among others, the use of recovery plans, multi-annual management plans and the possibility of applying emergency measures by the EU Fisheries Commission and single member states. However, one of the major problems in EU fisheries is the existent overcapacity of the EU fishing fleets in comparison to the available resource, and this has been handed back to the member states. As many EU fish stocks are being over-fished, the framework for the setting up of recovery plans to rebuild these fish stocks is a central element of the reformed policy. There are several good requirements which include:

- the respective stock has to recover within safe biological limits,
- the TACs and quotas shall be set on a multi-annual basis, within specific time frames,
- other stocks and fisheries have to be taken into account, and
- the recovery plans have to relate to other species and the marine environment.

On the down side, however, the final content of any recovery plan depends on the decisions made by the EU Fisheries Council and is, therefore, at risk of political bargaining and downscaling. In addition, a limitation of fishing effort is not obligatory.

The recovery plan for cod in the North Sea

The first recovery plan under the new CFP regulation has been adopted for the rebuilding of the cod stock in the North Sea. The objective is to ensure the safe recovery of the cod stock at least back to a target level of 150,000 tonnes. This level was defined as the precautionary level for the North Sea. Today, the spawning stock biomass is near 50,000, a third of the precautionary level and even less than the biological limit reference point of 70,000 tonnes.

Although the recovery plan claims to generally follow a long-term and precautionary approach, there are several drawbacks to this first recovery plan. Firstly, based on this plan, the TACs can only be changed by a maximum of 15% from one year to the next. In addition, the recovery plan concentrates only on cod, and furthermore, ICES has recommended zero catches of North Sea cod for the years 2003 to 2006. However, the EU Council has again agreed on a TAC of more than 20,000 tonnes for 2006. Thus, it seems quite unlikely that the cod stock in the North Sea will be able to recover in the near future, even with a recovery plan in place.

5 Conclusion

Overall, the examples touched on here show that fishery managers and responsible politicians are able to learn from the bad experiences of over-fished stocks in the past. However, it still remains an open question, as to how far the politicians are willing to translate this understanding into concrete measures. The demonstrated elements of a sustainable fishery will be able to safeguard fish stocks, fishermen's livelihood and the marine environment itself for the coming generations, but only if they are put into practise.

Literature

FAO (1995): Code of Conduct for Responsible Fisheries. Rome, 1995, www.fao.org/docrep/005/v9878e/v9878e00.HTM

FAO (2001): International Plan of Action to prevent, deter and eliminate illegal, unreported and unregulated fishing. Rome, FAO 2001, www.fao.org/figis/servlet/static?dom=org&xml=ipoa_IUU.xml

Hammer C, Zimmermann C (2003): Einfluss der Umsetzung der ICES-Fangempfehlungen auf den Zustand der Fischbestände seit Einführung des Vorsorgeansatzes (Influence of the implementation of the ICES advice on the state of fish stocks since the introduction of the precautionary approach). Inf Fischwirtsch Fischereiforsch 50,3:91–97

United Nations (1982): United Nations Convention on the Law of the Sea (UNCLOS), www.un.org/Depts/los/convention_agreements/texts/unclos/closindx.htm

Watson R, Pauly D (2001): Systematic distortions in world fisheries catch trends. Nature 414:2001:534–536

Oil Pollution in Marine Ecosystems – Policy, Fate, Effects and Response

Carlo van Bernem, Johanna B. Wesnigk, Michael Wunderlich, Susanne Adam, and Ulrich Callies

GKSS, Max-Planck-Straße 1, 21502 Geesthacht

1 Introduction

According to the article 21 of the Stockholm Declaration, following the 1972 UN Conference, all states have the responsibility to ensure that activities within their jurisdiction do not cause pollution in other states or in areas outside national jurisdiction. However, ever since the first transatlantic tanker ships left their launch-ways in the late 19th century, oil pollution has been a permanent danger for marine ecosystems.

There are generally three ways with which one can minimise the threat of oil pollution:

- The first is to increase the safety of navigation and the technical standard of ships
- The second is to increase the measures of control, and
- The third way is to amend the management respond measures to oil pollution.

Arrangements for the first point are mainly left to policy on the basis of national and international agreements. In this context, OILPOL (international convention for the prevention of pollution of the sea by oil, 1954) was the first global convention, which was organised by the International Maritime Organization of the United Nations (IMO). The purpose of this organisation is to establish a form of legislation in an almost lawless area, the open sea.

Article 1 of this convention summarises the tasks of IMO as follows: "to provide machinery for co-operation among governments in the field of governmental regulation and practices relating to technical matters of all kinds affecting shipping engaged in international trade; to encourage and facilitate the general adoption of the highest practicable standards in matters concerning maritime safety, efficiency of navigation and prevention and control of marine pollution from ships".

Aside from nautical safety and operational pollution concerns, the enormous growth in the amount of oil which is transported by sea has explicitly aggravated the danger of accidents and, as a result, the severe pollution of sea- and coastal areas.

It was the "Torrey Canyon" disaster in 1967 (about 120,000 tonnes of oil was spilled) which lead to the "International Convention for the Prevention of Pollution from Ships" (MARPOL 73/78), covering not only accidental and operational oil pollution but also pollution by chemicals, goods in packaged form, sewage, garbage and air pollution. Furthermore, this accident also showed that the ecological damage was caused not only by oil itself, but mainly by the chemical dispersants used to counteract the spill. Within MARPOL, the so-called Special Areas and Particular Sensitive Sea Areas (PSSA) are designated as areas with very high standards of control and even tighter restrictions for shipping.

Another disaster incident, the "Exxon Valdez" in 1989, where about 40,000 tonnes of oil befouled highly sensitive ecosystems, was the main reason why the US government established the "US Oil Pollution Act" in 1990 (OPA 90). This law not only initiated new technical standards (i.e. double hulls) but also the "Polluter Pays Principle" and further prevention measures with respect to poor-quality vessels. This law is enforced by the National Coast Guard and has resulted in a 95% decrease in the amount of oil spilled in US waters.

The controlling of technical ship standards and specifications (i.e. MARPOL) is basically the responsibility of the flag state and since the "Paris Memorandum of Understanding on Port State Control" (Paris MOU) in 1982 is now also under the control of the harbour state. The detection of oil on the sea is part of the overall tasks of the national coast guards using ships, aerial surveillance and satellites. For European waters, however, no common CG has existed until now. It needed two further oil accidents, one near the French coast (Erika, 1999, about 20,000 tonnes Heavy Fuel Oil) and another at the Spanish coast (Prestige (2002), about 70,000 tonnes HFO) to establish the European Maritime Safety Agency (EMSA). The task of this agency is to facilitate the prevention and the response to pollution by ships as well as to provide Member States and the European Commission with technical and scientific assistance (see Sect. 6).

The management measures undertaken to combat oil pollution is stipulated in national "contingency plans". These plans comprise the underlying strategies for a response, and depend on the characteristics of the coast in question and its sensitivity to oil pollution. It may be, e.g., that a fundamental decision to employ chemical surfactants (see Sect. 3) must be taken and then the potentials of bioremediation assessed. Aside of the methods and strategies of prevention and response to be used, the deployment of ships and equipment, the training of personnel, and the logistic evaluation as well as the use of drift models, are among the numerous aspects determined in these plans. Today, risk and consequence analyses as well as computer models are becoming of increasing interest as tools to assist in decisions on which strategies, methods and equipment (see Sects. 4 and 5) are the most appropriate.

Science, on the one hand, is looking towards to develop techniques for reliable and safe navigation as well as the for the safety of ships and is also used as a basis for the continuation of international/global conventions and agreements. On the other hand, other methods of prevention and response of oil pollution are urgently required to minimise the risk and damage to marine and coastal ecosystems. Although fewer in number, oil spill accidents are a permanent threat, therefore we will focus on subjects of the latter topic.

In this regard, research and development has to consider the fundamental behaviour of oil in the marine environment.

The fate of oil discharged into aquatic environments is very complex, and therefore only some of the more fundamental phenomena and processes can be presented here (see also Sect. 2). Using a strongly simplified classification, crude oil contains the compounds illustrated in Fig. 1:

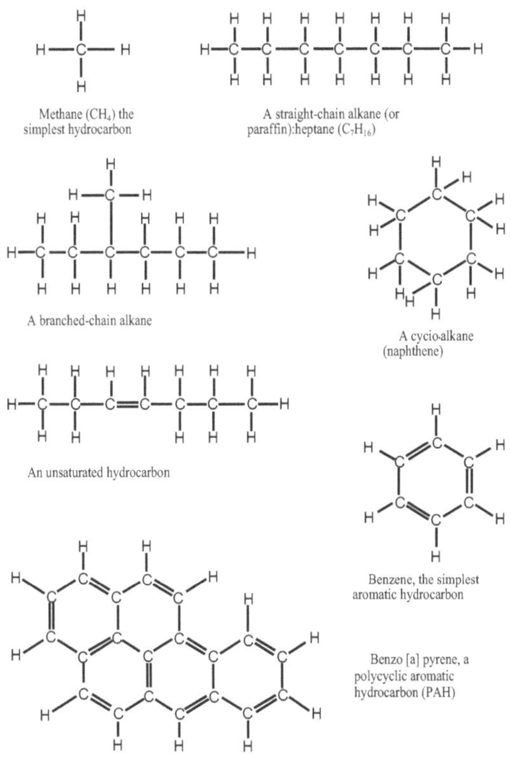

Fig. 1. The structure of some hydrocarbons (Clark, 1992)

In fact there are more than 10,000 components not including the substances added during, e.g., the refinery processes or modified in subsequent combustion or degradation processes. The components of crude oil are mostly homo-

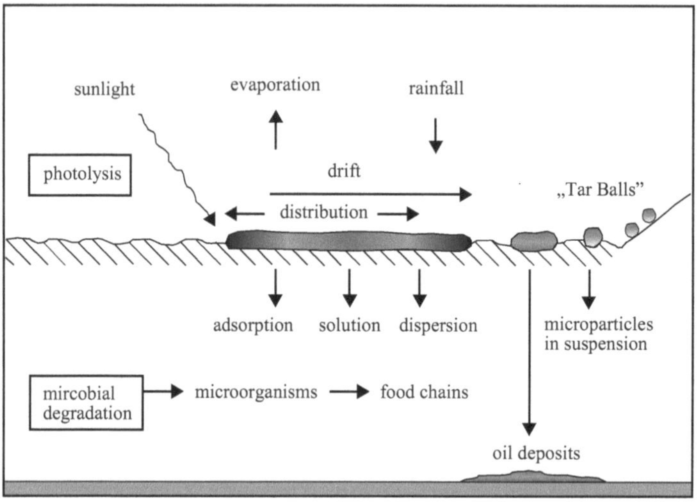

Fig. 2. Simplified patterns of oil spilled on water (Gunkel 1988)

logue chains of hydrocarbons and non-hydrocarbons (heterocyclic compounds like thiophene and pyridine and derivatives such as phenols). These oxygen, nitrogen and sulphur containing compounds usually form only a few percent of crude oil. The main compound classes are the straight or branched chains of aliphatics, the naphthenes (cyclic aliphatics with saturated rings), and, to a lesser extent, the aromatic compounds with one or more unsaturated rings. The composition of these compounds changes with each crude oil – there are not two oils being completely identical.

Because of this complexity and the different behaviour of the single substances after being discharged into the sea (different pathways of solubility, distribution, photolysis and microbial degradation) the chemical investigation of these subjects is among the most difficult and complicated tasks of organic chemists. To further complicate the situation, a wide range of hydrocarbons is of biogenic origin, produced by plants, algae and animals. They can be distinguished from petroleum hydrocarbons only by their composition (e.g. the predominance of odd chain length of n-alkanes).

The main processes involved when oil (especially crude oil) is spilled onto sea water, summarized as "weathering" (Dahling et al., 1990), are: spreading, evaporation, dissolution, formation of emulsions, dispersion in the water column, sedimentation and biodegradation (Fig. 2).

The direction and speed of a driven oil slick depends on the prevailing current conditions and about 3.5% of the wind velocity. Consequently, a drift model for coastal waters is a good tool to use as part of a conceptual model to predict areas at risk. A spreading slick itself forms a large area of "sheen", about 1 µm thick and containing less than 5% of the total oil volume. The majority of oil is bound to a much smaller area with a thickness of several millimetres in case of a stable emulsion. Within the first few hours or days most crude oils will lose up to 40% of their volume by evaporation. This process, which is driven by temperature and wind speed, reduces the proportion of lighter components in the oil, thus leaving a smaller volume with a higher viscosity and a reduced toxicity. This loss of oil components to the atmosphere is supplemented by a much smaller rate of dissolution. The amount of water soluble hydrocarbons around an oil slick is generally in the ppb range but remains toxic and bioavailable for marine organisms. On the other hand, the incorporation of water into the oil residue left by evaporation and dissolution leads to a large increase in the pollutant volume thus raising the viscosity once again. Very stable emulsions, as formed by some oils, are resistant to chemical treatments or heating. Under rough sea conditions, low viscous oils disperse naturally into the water column to a large extent, forming droplets of a wide range of sizes. While larger oil droplets resurface, only the smaller ($< 70\,\mu$) are found in permanent dispersion. Clay and particles of similar size (1–100 μm diameter) and microscopic organisms interact with dispersed oil droplets by adsorption and ingestion. In waters of high turbidity, as for example in estuaries, the resulting oil-mineral complexes can reach high levels and obey the characteristic environmental processes of sedimentation and accumulation in areas of low hydraulic energy such as near-shore tidal flats and watersheds.

The final weathering-process of spilled oil is biodegradation. All but the most refractory components of a crude oil can be degraded by biological action in the water column as well as in sediments. The rates depend on temperature and the availability of oxygen. They range from 1–50 mg m^{-3} day^{-1} to years in very cold or anaerobic environments.

How to combat oil pollution

Whenever a decision-maker has to decide which kind of action can be applied as a response to oil spilled at sea, a risk evaluation comes into play. Belluck et al. (1993) defined three classes of ecological risk assessment (scientific, regulatory, and planning) that lie along a continuum from most to least quantitative. Because cost (and usually time) increases with the level of scientific detail which can be obtained, the desire to improve analysis must always be weighted against the cost of the additional information.

With focus on the chemical dispersion of oil, the behaviour of oil in water should be assessed before the questions "Will dispersants work effectively with a particular oil in the environment of interest?" and "What are the ecological consequences of dispersant use?" can be answered.

The advantages of using chemical dispersants are twofold: in the first place they reduce the pollutant volume on the water surface. Secondly, they increase the rate of biodegradation processes by increasing the reactive surface of the oil. Their effectiveness depends mainly on the kind of the oil, its state of weathering (viscosity and degree of water-in-oil emulsions) and on the hydraulic energy in the area of concern. Other factors of gradual influence include: salinity, turbidity and temperature.

There is no conclusive way to estimate their usefulness in a particular environment with varying conditions. Appropriate scenarios for individual cases have to be evaluated during a comparative analysis of risks and benefits.

A fundamental goal for any oil spill response is to minimize the ecological impacts of a spill (Lindstedt-Siva, 1991). The decision as to whether it is better to protect sensitive habitats – rather than to optimize cleanup –, requires a specific methodology to optimize all possibilities of response into an integrated program. In this regard a "Net Environmental Benefit Analysis", NEBA, (Baker, 1995) based on an ecological risk assessment approach can serve as part of the integrated precaution measures.

The activities involved in the assessment can be summarized in three phases: problem formulation, analysis, and risk characterization (US EPA, 1992a). Within these phases, quantitative as well as qualitative data may be used depending on the state of knowledge about the systems involved. The uncertainty of data and methods has to be defined as far as possible before the resulting information can be incorporated into conceptual or mathematical models.

In the following J. B. Wesnigk summarizes some fundamental aspects of source and fate of oil in the sea; M. Wunderlich describes the effects of chemical dispersion and S. Adam and U. Callies present examples to model risks and consequences of oil spills.

Following the accident of the oil tanker ERIKA in December 1999 off the French coast, the European Parliament and the Council adopted regulations which established the European Maritime Safety Agency (EMSA), a first step to establish a common European agency for the response of marine pollution. B. Bluhm focuses on the legal basis, its approach and the activities of this agency.

2 Fate and Effect of Oil Pollution in the Sea

Johanna B. Wesnigk

Human activities can have major effects on the marine and coastal environment. In order to prevent and be prepared for accidents several instruments have been devised to avoid the worst effects of pollution and other destructive influences, for example:

- Environmental impact assessment – to be used in the planning stage for major projects
- Environmental management – to be used during the operation of facilities which can have major effects on the environment

- Contingency planning – this can be part of environmental management or – as in the case of the protection of coastal states' waters from shipping and offshore accidents – standing alone. It should include consideration of biological aspects, i.e. the sensitivity of coastal and marine organisms and resources.

Sources of marine oil pollution

Table 1. Sources for 1999 amounts and percentages: Fairplay International Shipping Weekly, London, 30 October 1997; Alexander's Gas & Oil Connections Reports, Vol # 3, Issue 22, December 1998; Fritjof Nansen Institute, Norway. 1994; Smithsonian Institution – Website of the Ocean Planet Project, 1999

Category	Sources	% 1999	% 2002
DOWN THE DRAIN	used engine oil oil runoff and wastes from land-based industrial sources other land-based sources	53	37
ROUTINE SHIP MAINTENANCE AND OPERATION	bilge and tank cleaning fuel leakages oily ballast water oily and wastes cargo residues	20	incl. above
UP in Smoke	atmospheric deposition of hydrocarbons in the seas through air pollution from vehicles and industry	13	–*
NATURAL SEEPS	oil released from the ocean bottom and eroding sedimentary rocks	9	46
Big SPILLS	tanker accidents accidents of other types of ships	5	12
TOTAL 1999 = 2.6 mio t TOTAL 2002= 1.3 mio t			100%

*no estimate in the US study (see below), add 3% for discharges due to extraction of oil

Table 2. Simple Classification scheme of hydrocarbons for the determination of combating options and biological effects. (O'Sullivan & Jacques 2001)

Type of oil	Volatility	Solubility in Water	Natural dispersion	Response to dispersants	Stickiness	Biological harmfulness
I Light volatile	High	High	Disperses easily	Responds very well	Not sticky	Highly toxic
II Moderate – heavy	Up to 50% can evaporate	Moderate	Some components disperse	Responds easily	slightly to moderately sticky	variable toxicity
III Heavy oils	¡ 20% can evaporate	Low	Little dispersion	with difficulty	very sticky	Smothering, clogging
IV Residual	Non-volatile	very low	no dispersion	not at all	very sticky to solid	Smothering, low toxicity

In 2002, the National Research Council (NRC) of the U.S. National Academy of Sciences has estimated the average total worldwide annual release of petroleum (oils) from all known sources to the sea at 1.3 million tonnes. However, the range from different sources is wide, from a possible 470,000 tonnes to a possible 8.4 million tonnes per year.

Biological effects of oil

Three groups of factors are important to characterise any spill:

1. Circumstances of the spill
2. Factors modifying the spill behaviour
3. Factors influencing the impact of the spill on marine organisms and amenities.

The effects of oil on marine life can be considered as being caused by either its physical nature, i.e. physical contamination leading to clogging, interference with mobility or feeding, or smothering, leading to suffocation, and heat stress.

On the other hand biological effects can be dominated by the chemical components of the oil, leading to:

1. Acute toxicity – lethal, sublethal, immediate effects
2. Chronic toxicity – delayed effects
3. Bioaccumulation, especially in molluscs like mussels
4. Tainting of sea food

The ability of animal and plant populations to recover from an oil spill depends on the following characteristics: Abundant organisms with highly mobile young stages which are produced regularly in large numbers may repopulate a cleaned-up area rapidly. Slow maturing, long-lived species with low reproductive rates may take many years to recover their numbers and age structure. Examples for the latter group are: seals, otters, reptiles (turtles), whales and dolphins and some cold water fish. In general, the rate of recovery in tropical or Mediterranean regions is faster than in cold environments like the North Atlantic and the Baltic Sea.

The main factors affecting the rate of biodegradation, one of the few biological processes leading to the disappearance of oil components, are

- Temperature
- Availability of oxygen, and
- Availability of nutrients, mainly nitrogen and phosphorus.

In the best case oil components – mostly linear chain and small aromatic structures – are degraded to carbon dioxide and water; a biotransformation into potentially more toxic components is also possible. This process is only efficient with relatively small amounts of oil and near optimum conditions. Otherwise, either the time span is very long (years to decades) or a strong development of oil degrading microbes can cause oxygen deficits and accumulation of biomass (slime, mats etc.).

Integration of biological factors into oil spill combating procedures

A combination of Contingency Planning and Sensitivity Mapping, two useful tools to deal with all the information in a systematic and operational way, helps to prepare for a fast and efficient reaction in case of major oil spills.

It is, therefore, very worthwhile to do systematic mapping of the biological resources and coastal features before a major accident takes place. Such data can greatly support future decisions which can then be taken on a sound scientific basis, and compensation claims and claims to recover clean-up costs are substantially facilitated by good background data.

Decisions concerning the priorities to be put on sensitivity maps should be taken in cooperation with all relevant stakeholders. It will be hard to achieve consensus, but it is crucial to reach the best possible agreement before a spill has occurred.

The simplest approach to sensitivity mapping is to follow an Environmental Sensitivity Index (ESI) approach, which is based on mapping the physical and geological features of coastlines. Gundlach & Hayes (1978) have developed a priority list of vulnerability or sensitivity for different coastal habitats against oil. Their concept is based on geomorphological parameters. It was designed for a wide array of different coastal regions, taking into account an array of sedimentological, biological, meteorological and hydrographic factors. Later on it was slightly modified to the ESI – Environmental Sensitivity Index (Michel & Dahlin 1993) which grouped ecosystems into five sensitivity groups. A full description of categories ESI 1 (e.g. exposed rocky headlands) to the most sensitive category ESI 10 (e.g. salt marshes) can be found in IMO/IPIECA Report on Sensitivity Mapping for Oil Spill Response which also includes pictures of the different types of coastline. A more recent development is the European Impact Reference System (IRS). This IRS uses the same basis as the old ESI classifications but goes into more detail with regard to ecosystem values and functions.

The overall sensitivity of an ecosystem to oil pollution is determined by:

- The vulnerability of the habitat or physical environment
- The sensitivity of the populations or communities of the organisms living in that environment
- The resilience of the living community as a whole – called recovery

For most sensitive areas the seasons play an important role in the exact determination of the sensitivity and combating options, for example because bird colonies might be populated or not, birds might be moulting, migratory birds might be feeding etc. at certain periods of the year (van Bernem et al., 1989 and 2000).

An example of the risk posed to a sensitive area can be found in Sect. 5, using the inner Jade Bay in the German Wadden Sea as an example.

Acknowledgements

Some of the material used in this article was compiled within the EC-funded project BIOREM (Asia Pro Eco programme) Bioremediation of oil spills in Vietnam. The results of the project, available free of charge for downloading, include Guidelines for Bioremediation of oil spills in Vietnam, Training material for Bioremediation of oil spills and Presentations of the workshop and training course held in 2005 can be found on the website: http://www.hs-bremen.de/ikrw/biorem/index.htm.

3 Chemical Oil-spill Countermeasures at Sea – Dispersants

Michael Wunderlich

Spraying chemical dispersants can be an efficient tool of controlling oil slicks and may be an alternative to other combating methods such as mechanical clean-up, shoreline cleaning, burning, bioremediation or “leave alone” (monitoring only).

A dispersion is a colloid distribution of an insoluble or barely soluble liquid (dispersed phase, droplet size 1–50 μ m) in another liquid (closed phase), for example oil in water.

Oil-spill dispersants are mainly composed of two basic groups of chemicals. These are surface-active agents (surfactants) and solvents. Modern products are so-called concentrates with 25 to 60% surfactants and polar organic solvents or hydrocarbon solvents. Due to their molecular characteristics, dispersants have an affinity to oil as well as to water. They settle at the interfacial

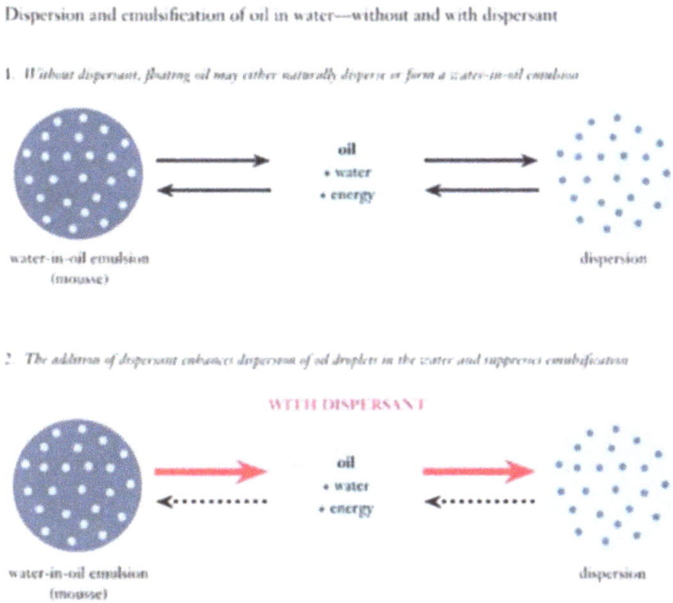

Fig. 3. Dispersion and emulsification of oil in water (IPIECA, 2001)

surface between oil and water. By the input of mixing energy oil-slicks break up into droplets. In the presence of dispersing agents, these droplets form easily with minimum energy requirement and build a stable dispersion (Fig. 3).

Dispersants may be sprayed from aircraft (fixed-wing planes or helicopters) or from ships. Their application is done most effectively against the wind. Smoke bombs are used to indicate the direction of the wind. Extremely thin oil layers appear as silvery sheen or as rainbow colours on the water. These types of slicks need not be combated; they will disappear after a short time by natural weathering. Therefore, the strike teams must only spray dispersants onto the thick-layered parts of slicks. Weathered oil, or "chocolate mousse", is a water-in-oil-emulsion. This kind of emulsion is hardly affected by dispersants.

When and where to use?

Favourable conditions are:

- Fresh oil not exceeding a viscosity of 3.000 mPas;
- The colour of the oil slick should be dark grey to brown (equivalent to a thickness of more than 100 µ m);
- The sea should not be completely calm (sea state > 0) to guarantee a minimum of mixing energy;
- Strong currents cause rapid mixing and dilution of the oil;
- The wind speed should not be below Bft 1 and not exceed Bft 6;

- When oil is moving towards the shore and all possible response methods are required to minimize the environmental impact;
- When physical removal methods alone are not adequate;
- When weather and sea conditions preclude the use of mechanical recovery systems;
- When the natural dispersion is not fast enough;
- When birds occur in large numbers in the area (Page, D.S. et al., 1983).

Dispersants must not be used:

- In shallow water with poor circulation;
- In the catchment areas of seawater desalination plants;
- On heavy oil, high-viscous oil and "chocolate mousse";
- When the water temperature drops below the oils pour-point;
- Near seagrass beds;
- Near fish farms and mariculture facilities or aquarium intakes.

Information about the dispersability can be derived from knowledge of the physical and chemical properties of the spilled oil and the ambient temperature of the seawater, by a screening test, and by monitoring the results of the first run of the dispersant application.

Dispersed oil has a milky appearance. This shows that the dispersant is efficiently working. Under field conditions, a spray path can be clearly differentiated from the polluted sea area.

Advantages of dispersant application

Dispersants can be used in a wide range of weather and sea conditions. It is usually the quickest response action.

Shoreline contamination and the risk of fire hazard are minimized, and the contamination of birds and sea-mammals can be reduced. Dispersion also improves the natural biodegradation of the oil in the environment, and often the formation of "chocolate mousse" is inhibited. In most cases the costs are much lower compared to mechanical combating.

Disadvantages of dispersant application

The weathering of oil causes changes its characteristics. For example, the viscosity of the oil can rise quite rapidly. This leads to a loss of dispersant efficiency. The time span when combating is possible is called a "window of opportunity". Usually this "window of opportunity" is open only for a short period of time (hours). Not all types of oil are dispersible; for example, heavy fuel oils can barely be affected by such dispersants. Dispersed oil may be harmful to benthic organisms in shallow waters and can remain in the water column. Malfunctions of installations may also occur and, finally, the input of chemicals into the environment also means additional pollution.

Ecological considerations

Oil is removed from the sea surface by dispersing. This means seabirds and sea mammals are protected from contamination. Moreover, the pollution of shorelines can be reduced and thus shoreline-dwelling organisms are also protected. In the open sea, however, plankton in the upper water layer (0-10 m) is affected by oil pollution. The higher mobility of oil droplets in the water body can also have an impact on bottom-living or benthic organisms. Many of these organisms are important food sources for fish. Tidal mud-flats and sea-grass areas, coral reefs and mangrove forests are very sensitive habitats and take years to recover from a heavy pollution. In order to evaluate all ecological aspects, a so-called Net Environmental Benefit Analysis may be undertaken. This can provide the basis for deciding whether to use dispersants or not.

Decision making

There are numerous methods which assist the decision making processes with regard to chemical oil-spill response. For example: Operational Response Guidelines (Response Manual, CEDRE, 2005)

When dispersant use is the preferred choice, information is required concerning the following aspects: Spill volume, layer thickness, physical form of the spill, type of oil spilled, spill location, weather, sea state, and water currents. These criteria are the basis for computer-aided decision making, especially the ranking of combating methods (Operational Response Guidelines, RIJSWIJK, 2003).

Tests and approval

Dispersants are tested usually for effectiveness and toxicity. Laboratory methods for effectiveness are: Mackay test, EPA test (SET), Labofina test, Mackay flume test, Swirling flask test, Paddle mixer test, EXDET. None of these tests is fully satisfying. Oil-Water-Ratio, Dispersant-Oil-Ratio and energy rating differ widely. One can compare different products only under defined test conditions.

Toxicity tests. Different test organisms are used as representatives of the fauna and flora present. Typical representatives include species of planktonic algae, crustaceans, mussels, and fish. One criterion is the lethal concentration at which half the number of the test organisms die within 24 h or 48 h (Gilfilan, E.S. et al. (1983, 1984)).

Table 3. Characteristics of the U.K. toxicity test

Test species	Criterion	Test carried out	Oil
Shrimps (Crangon crangon)	Death rate	1. Crude oil alone (sea and beach test)	Fresh Kuwait crude oil
		2. Oil plus dispersant (sea test)	
Common limpet		3. Dispersant alone (beach test)	

Approval. Dispersants are used in many countries, including the U.K., Korea, USA (many coastal states), Australia, Egypt, France, Greece, Indonesia, Italy, Japan, Malaysia, Norway, Singapore, Spain, Thailand, and most of the coastal African, South American, and Middle East countries.

Due to a restrictive-use policy, there are no stockpiles of dispersants and no officially tested or approved products in Germany. In the case of a severe oil pollution within the territorial waters of the North Sea, German authorities can ask the members of the Bonn Agreement for assistance. This may include the request for dispersants.

National and international regulations

In Europe, there are agreements of mutual assistance in cases of severe oil pollution of the North Sea, the Baltic Sea, and the Mediterranean Sea. Germany has signed similar bilateral agreements with the Netherlands and Denmark. An agreement with Poland is in preparation. An important national law in this context is the Federal Water Act (Wasserhaushaltsgesetz). This is the frame for legal regulations of the Länder (federal states), and it stipulates that any kind of water use must have permission by the pertinent authorities. This is also valid for the application of dispersants.

Case studies and documented incidents

Case studies and incidents describe the efficiency of dispersants under real-life conditions. Examples of important case studies include:

- 1997 North Sea trials, Location: East coast of U.K., Lowestoft area. Oil type and quantity: Alaska North Slope Crude (ANS), 30 t. North Sea Forties Blend, 2 × 50 t. Dispersant: COREXIT 9500 and Dasic Slickgone NS. Application by aircraft, The conditions were: ANS weathered for 55 h; the emulsion viscosity was between 15.000–20.000 cP (wind: 8–10 kn; Tw = 15 °C), the emulsion had a water content of about 30 Vol.%.
  Results: the emulsion broke up soon after application of the dispersant. The dispersion was monitored under the sea surface.
  The oil type North Sea Forties Blend was weathered for 44 h. The emulsion had a viscosity of ~ 4.500 cP. The dispersant COREXIT 9500 was sprayed

during trial No.1, and the product Dasic Slickgone NS was sprayed during trial No.2: A rapid reduction of thick parts of the slick was observed. Only broken sheens remained after 14–18 hours.

- Bunker No. 5 fuel oil (IFO-180) weathered for 4 h and 24 h was used in another test. The viscosity ranged from 8,000 cP to 10,000 and 23.000 cP. The results of the first application showed 50–75% dispersion, while the second application was less effective. Tar balls remained at the surface.
- The "Sea Empress" incident. The "Sea Empress" spill off the coast of Wales, U.K. in 1996 is well documented. 72,000 t Forties Blend crude were released over a four-day period. In total 445 t of dispersants and 8 t of demulsifier were used between 18–22 February. On 25 February, a slick of heavy fuel oil was sprayed with 1.5 t COREXIT 9500. This run failed to disperse the slick. The application from airplane was guided by a remote-sensing aircraft and a ground team monitored the effects on the oil. Because of the missing efficiency in this case, the use of dispersants was restricted and spraying was not allowed nearer than one kilometre to the coast, thus guaranteeing a minimum water depth of 20 m.

Table 4. The "Sea Empress" incident: Comparison with other spills (Sea Empress Environmental Evaluation Committee (SEEEG): Initial Report, July 1996; The Sea Empress incident. Report by the Marine Pollution Control Unit. The Coast Guard Agency, Southampton, 1996)

Oil fate	Exxon Valdez	Braer	Sea Empress
Spilled T	37,000	84,000	72,000
Evaporated %	20–30	9–19	35–45
Dispersed%*	20–25	46–56	45–59

*dispersed naturally and by chemical treatment

4 Oil Spill Response Management Supported by Models and Integrated Software Systems

Susanne Adam

AMOCO CADIZ, ERIKA, and PRESTIGE - these and other ship names not only stand for huge oil spills but also for disastrous damage to coastal environments and regional economies. These accidents have a common history in that the authorities were unprepared for such an emergency and political confusion and uncertainties existed in the decision-making processes. Given the extent of these disasters – each with spilling 10 million-plus gallons of oil into the water – chaos is not surprising. The previous chapters gave a first impression of how complex the spreading and weathering processes are when oil is discharged into the sea. These processes must be considered in conjunction with the available knowledge on ecological and socio-economic sources before any decision is taken concerning the appropriate combating measures to use. Computer models and software systems help in providing an objective information platform when dealing with such kinds of complex decision making problems and thereby help in providing efficient response scenarios to marine pollution emergencies. A wide variety of different systems has evolved over the past few decades and range from simple models forecasting the drifting of an oil slick to integrated models which combine, for example, information on drift, weathering and environmental impacts to even more complex and sophisticated Decision Support Systems,. In the following, an overview of the purpose and structure of oil spill models and software systems is given. State-of-the-art science and technology is highlighted as well as areas of active research. Finally, an example reconsidering the Prestige accident shows the potential of analyzing complex spill events and outlines how to arrive at decisions which minimise the economic and environmental impact by using Decision Support Systems.

Purposes and Objectives of Oil Spill Systems

Nowadays all maritime countries agree on the necessity of putting in place computerised tools to better respond to maritime pollution emergencies (Leito, 2003). That includes not only tools applicable to the emergency case itself but also systems aiming to provide effective emergency preparedness. The degree of success in clean up and spill management is strongly dependent on such early warning and development planning. Software systems can be categorised with regard to their threefold purpose: (1) Planning and simulation, (2) Operational monitoring and (3) Crisis management. Systems which belong to the first category aim towards the planning of response strategies and purchase of contingency equipment. This includes both national emergency response plans and resource planning for offshore installations. The response techniques and strategies employed in a spill depend on the product spilled, its quantity, the location, the response times, weather conditions, responder capability and the availability of response equipment. Thus, a significant number of simulations with varying input parameters have to be carried out in order to find the optimal response strategies for a wide range of possible oil spill scenarios under different conditions. Further applications can involve the training of decision makers and risk assessments of offshore installations.

Early awareness of the potential risk of an average oil spill can become crucial in order to avoid potentially disastrous impacts. Today, new operational monitoring techniques such as radar remote sensing, satellite imagery and other visual techniques used, for instance, in conjunction with aerial observation allows for an early detection of spills in the marine environment (Mansor, 2002). When such techniques are integrated into automatic warning systems, the response time can be reduced quite significantly. Furthermore, systems which monitor maritime traffic are the best means to detect disabled

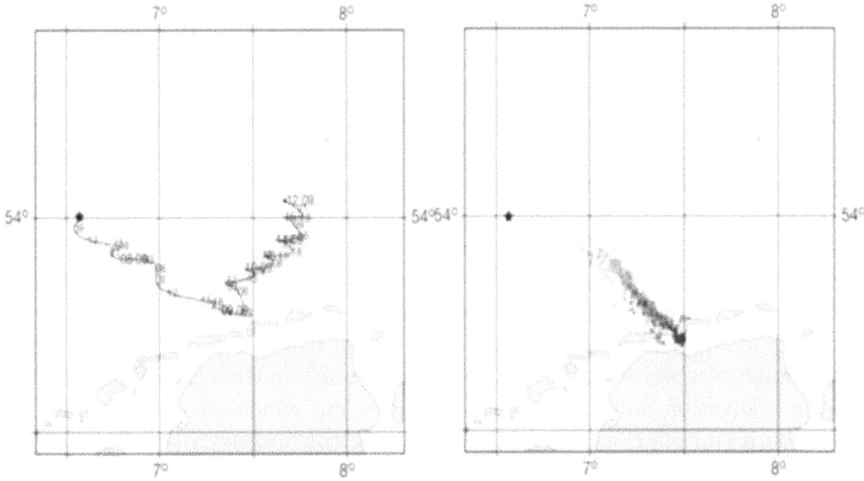

Fig. 4. Oil drift and dispersion simulation with the integrated model system of the Federal Maritime and Hydrographic Agency (BSH) Germany. The left figure shows the hypothetical drift of 1000 tons of heavy bunker oil released north the isle of Borkum, German Bight, on 7th September 2001. The right figure shows the distribution of different components two days after release. ×: Oil at the surface, +: Dispersed oil, ∘: Oil at the sea bottom. Evaporated oil and water content of surface oil is calculated but not shown

ships at an early stage to enable appropriate action to be taken in order to avert an emergency situation.

Once an oil spill occurred, systems which belong to the third category are suitable for facilitating a fast decision making process. Mathematical models and software systems are capable of answering a wide range of management questions which may arise during a crisis situation. Forecasting drift, spreading and weathering of an oil slick is the most fundamental capability of todays applied systems. However, state-of-the-art developments can also assess the environmental impact or the success of the response measures, and may even be able to combine both aspects. Furthermore, significant scientific effort is being given to the development of multi-criteria decision support systems, which would be in a position to suggest the most reasonable response strategies in terms of the various ecological and socio-economic criteria present (Wirtz, 2004).

Design and Structure of Oil Spill Systems

As many objectives can be covered as many different computerised systems were launched. Also each country, state, local authority or oil company, who needs to be prepared for a marine pollution emergency, applies various systems with varying complexity depending on the risk potential, the local conditions, the frequency of use or even on the acceptance of software systems to hand. In the simplest case, a mathematic model is used to forecast oil spill trajectories to indicate threatened coastlines. Such models also form the kernel of more complex systems since their results are crucial for further conclusions regarding contingency planning and protection of sensitive ecological and important economic areas. Integrated model systems are the most advanced systems in use today. As the name suggests, different models are coupled in order to predict, for instance, and in addition to the trajectory, the weathering of oil or its immediate toxicological effect on organisms. Examples of the results of an integrated model system are shown in Fig. 4. By the addition of evaluation methods or a decision support module with which, for example, different containment strategies can be compared, a system can then be defined or classified as a Decision Support System. Such systems which are often coupled to spatial data can offer far more support during contingency planning and the protection of sensitive ecological and important economic areas.

From experience, integrated model systems and Decision Support Systems are highly complex and consist of numerous modules and interdependencies. However, a general structure can be derived, namely: Databases, often embedded in Geographic Information Systems (GIS), provide the basis for each system and he provided information will flow into a structure of coupled numerical and/or stochastic models. Finally, the results are presented by means of a user interface, often accessible again by a GIS.

State-of-the-Art Examples: OILMAP & OSCAR

Despite the wide range of software systems used for oil spill management and response purposes, only a few systems are available which allow for comprehensive and reliable assessments and which then make these assessments easily accessible in a user-friendly manner. In the following, two state-of-the-art examples for integrated model systems are described.

Applied Science Associates, Inc. (ASA), USA, developed the OILMAP - Oil Spill Model and Response System. The system can generate rapid predictions of the movement of spilled oil. It includes simple graphical procedures for entering both wind and hydrodynamic data as well as for specifying spill scenarios. OILMAP is designed to track and predict the spatio-temporal distribution of surface oil and to assist in spill response efforts and training. Due to the wide range of functions, many companies, consultants, research institutes and governments world-wide use OILMAP (Zigic, 2004). The following five mathematic models are integrated in OILMAP: Trajectory and Weathering Model, Stochastic Model (for Risk Assessment and Oil Spill Contingency Planning), Receptor Model (for Oil Vulnerability Analysis), Search and Rescue Model, and Subsurface Transport Model. The system incorporates further an embedded GIS for storing any type of geographically referenced data and can be used to display model predictions in relation to important or sensitive resources.

The unique stochastic model feature of OILMAP should be pointed out here. This model can be used to determine the range of distances and directions which an oil spill from a particular site is likely to travel, given the historical wind speed and direction data for the area. With help of the stochastic model, a large number of single simulations for a given spill site can be performed by varying the wind conditions for each scenario. The trajectories are then used to generate probabilities as to whether surface waters or shoreline areas will be oiled. Practical use of such information includes also the determination about where response equipment is best positioned in order to be most effective (Zigic, 2004).

Another state-of-the-art software includes the OSCAR - Oil Spill Contingency And Response model system which was developed by SINTEF of Norway. The OSCAR model system was developed as a tool for providing an objective analysis of alternative spill response strategies. OSCAR is primarily intended to help achieve a balance between the costs of preparedness in the form of available, maintained spill response capability on the one hand, and potential environmental impacts on the other. Thus, this system serves mainly the needs of the petroleum industry, the regulatory industry, and the marine insurance industry.

The key components of the system comprise an empirical oil weathering model, a three-dimensional oil trajectory and chemical fates model, an oil spill combat model, and exposure models for fish and ichthyoplankton, birds, and marine mammals. Control of simulations as well as a visualisation of results is facilitated by a graphical user interface, which also offers access to a variety of databases and tools. With the latter, a user can, for example, create or import wind time series, current fields, and grids of arbitrary spatial resolution (Aamo, 1997).

In order to improve the most standard of integrated oil spill systems, a multitude of combat activities including mechanical (skimmer/ boom systems) and chemical response (dispersant application) are simulated in a realistic way using a multi-agent approach embedded in the oil spill combat model. By providing a quantitative and objective assessment of each contingency performance, an optimal response can be proposed based on a set of physical or biological criteria for the potentially affected area. Expenditures, effectiveness (as defined on the basis of various criteria) and the environmental benefits of various options can be compared. Accordingly, it can then be decided which response method (mechanical containment/ recovery, dispersant use) is the most suitable way of dealing with potential oil spills.

However, both the OILMAP and OSCAR systems leave the interpretation of the amount of pure oil stranded on different coastal or offshore areas up to the user. One unit of oil can have rather different consequences for different ecosystem functions or economic resource uses. Hence, the lack of any post-processing of the different complex impact scenarios limits the extent to which a systematic comparison of response options can be made.

The Prestige Spill off the Spanish Coast

In November 2002, the Prestige oil tanker hit rough weather off the coast of Spain and was badly damaged; the crew air-lifted to safety and the vessel was led out to the sea. After 6 days moving and spreading out oil along the whole Galician coast the tanker was torn in two on November 19th at the southern edge of the Galician Bank, 133 nautical miles from the Galician coast in an area with depths of about 3500 m. The thick black liquid – it was estimated that about 80% of the cargo 78,000 tons in total spilled out – spread over coastlines and wildlife, prompting fears that the environmental disaster could be even worse than the one in Alaska caused by the Exxon Valdez 13 years earlier. Finally, the Prestige incident rose to a symbol for a tremendous ecological disaster and political failure.

By proposing, in the following section, a method as to how the decision making procedure could have been improved by the use of sophisticated Decision Support Systems, does not imply that the Spanish scientists did not try to support their government. Indeed, they did this by, for example, the provision of daily predictions of the trajectories (Leito, 2003). Given the well-known winter climatology, the decision to move the vessel offshore to the southwest was a consequence of poor communication between the government officials dealing with the spill and the scientific and technical communities, rather than a deficit of knowledge. The result of this unfortunate decision was the spreading of oil across more than 1000 km of shoreline (Serret, 2003).

Since the affected area included a large number of protected habitats which make up the NATURA 2000 network of the EU, including the Costa da Morte estuaries and lagoons, the environmental damage was enormous. Rich tidal zones and wetlands which form part of the RAMSAR protected habitats were dramatically affected as was the National Park of the Atlantic Islands, home of the four most important colonies of marine birds in the Atlantic areas (ETCOTE 2003). However, these ecosystems have also a high social and economic value. Northwest Spain is the main reservoir of seafood in Europe. Fishermen and the fish industry are the most affected in a society which depends mainly on the sea for its living. Moreover, the tourism sector, which is based mainly on rural and coastal tourism, is expected to suffer for years to come. The long-term economic damage, including the response measures and cleanup, was calculated to be 10 billion Euros (Clauss, 2003).

The Prestige Disaster Revised

Given the above facts it must be asked how the ecological and economic impacts of such an oil spill could have been reduced. By a combination of modelling and evaluation methods with which different containment strategies can be compared with respect to a variety of potential ecological and economic damages, the idea of the Spanish government to tow the Prestige vessel in different directions will be re-addressed here. In particular, the above described OSCAR system coupled together with multi-criteria analysis to a

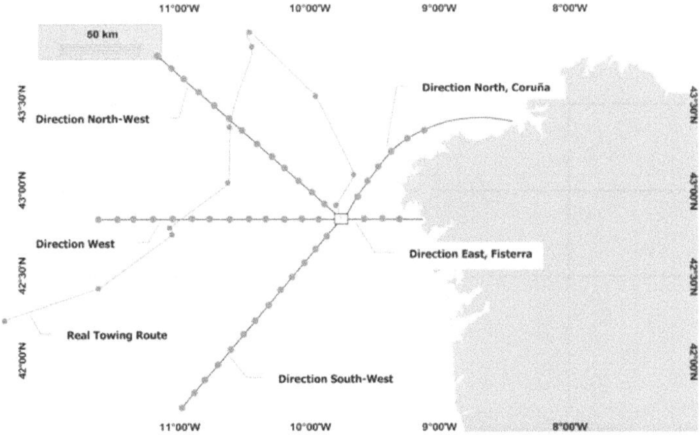

Fig. 5. Towing routes and hypothetical sinking positions after 6-hour towing intervals (*grey points*). It is assumed that the Prestige is dragged at the rate of 32.5 km per day for all directions. The original towing route is shown in the background as zigzag course

multilevel Decision Support System was used to assess impacts of five different towing directions. In order to emulate the operation conditions during the accident as precisely as possible, only those weather and current forecasts as well as the contingency equipment available at that time of the accident were incorporated into the analyses (Wirtz, 2004).

The five different towing directions correspond to securing the vessel in a nearby harbour or a non-exposed bight against dragging it into the open sea to deliberately sink it. Although all dragging directions were an option at that time, the uncertainty regarding, for example, the imminent break-up of the ship seemed to render any rationale judgment or assessment of the resulting consequences nearly impossible. In this case study, the five towing directions were defined as follows: northwest, west, southwest, east to the bight of Fisterra, and north to the harbour of Coruña 3 to 13 potential grounding locations were distributed along each towing route, and separated by a maximum three day trawling distance (Wirtz, 2005). This yielded a total of 49 different positions as shown in Fig. 5.

For each potential grounding location the spatial and temporal evolution of the oil slick was hindcasted using OSCAR - Oil Spill Contingency And Response System. Protected nature areas as well as areas of economic valuable were considered within the model system with the help of a GIS. The pollution intensity at each area was recorded and different methods applied to transform tons of oil into impacts by: (1) linear model (2) fuzzified function (3) economic evaluation (Liu, 2004). In a second step, a sinking probability was attached to each grounding scenario depending on the estimated time when the ship would break apart. As a last step, when this information was coupled to a

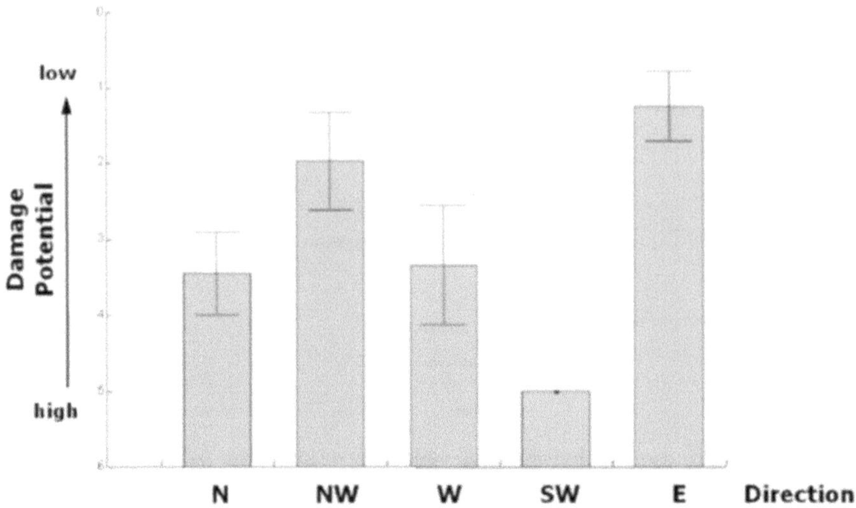

Fig. 6. Mean ranking of the five possible response measures which included the sinking uncertainty, the socio-economic and the ecological criteria. Standard deviations are derived from variations in other insecure parameters

multi-criteria analysis (MCA) module an objective decision could be achieved against a set of ecological and socio-economic criteria. The economic losses were assessed on the basis of income loss to fishery, mariculture, tourism and transport. The income loss was in turn estimated using sub-regional yearly data of the amount of fish landed, mussels harvested, cargo cleared and beds occupied. Ecological impacts were differentiated with respect to three semi-quantitative indicators of the sensitivity and importance of the local habitats, and included: reproductive capacity, vulnerability and protection level.

A final ranking of the different measures results from aggregating the compiled criteria factors in a politically balanced way. The ranking of the five possible towing directions, including also the sinking uncertainty, (see Fig. 6) reveals that the northern and eastern towing directions are clearly favoured. Furthermore, the easterly direction, which corresponded to the option of securing the vessel in the protected bight near Fisterra, clearly stands out as the optimal solution. The estimated overall damage is highest when the vessel is brought in a south-westerly direction. Thus, for the ocean-going scenarios, the north-westerly direction, which indeed was the first choice of the authorities, would have provided the highest probability of keeping the environmental and socio-economic impact on Spanish coastal areas at a minimum, since most of the oil would have drifted to the Gulf of Biscay.

5 Long Term Simulations of Environmental Conditions as a Basis for Oil Spill Risk Assessments

Ulrich Callies

Environmental damage to specific coastal regions by a given oil spill depend crucially on the prevailing winds, waves, temperatures etc.. Some of these parameters are also relevant for the risk that the accident giving rise to an oil spill takes place, i.e. the probability of a certain type of accident and the probability of disastrous environmental impacts must not be considered as independent modules in the context of an overall risk assessment. It is a common approach to take into account the variability of weather parameters by statistical representations of their observed frequency distributions. However, winds and waves, for instance, are not independent of each other. In addition, the temporal evolution of weather patterns exhibits certain time scales, which may become important when periods of several days are investigated. Typical transition probabilities between different weather patterns also exist. However, it is impossible to represent such co-variations of different variables in time and space in terms of simple statistical summaries. Instead, using all the information available about past environmental conditions ensures that less frequent unfavourable scenarios will also be properly dealt with.

The long term environmental data set

All example calculations reported in this chapter make use of a multi-purpose data set which has been generated to achieve the best possible representation of conditions over the past few decades, and comprises consistent information about the winds, North Sea currents and waves. Wind analyses are based on operational reanalyses (Kistler et al., 2001) of the National Centers for the Environmental Prediction (NCEP) covering the whole globe and with a spatial resolution of about 200 km and a temporal resolution of 6 hours. The analysis begins by running a state-of-the-art numerical weather model including the assimilation of observations. Dynamical downscaling was applied to the NCEP data to improve their resolution (50 km) for regional applications over Europe (Feser et al., 2001; Meinke et al., 2004). These atmospheric data were then used as wind forcing for both a hydrodynamic model of the North Sea (Weiße and Plüß, 2005) and a wave model. For several decades (1958-2002) the combined model results were stored on an hourly basis. Pre-calculated time-dependent fields of wind, waves and currents allow for a large number of different oil spill simulations without the necessity of computationally demanding hydrodynamic simulations. To do otherwise, it would be impossible to perform the many parameter variations, including, for example, variations of the location of the assumed accident or variations of the oil characteristics, needed for a comprehensive risk assessment.

Applications

Figure 7 illustrates the first step towards an analysis of the vulnerability of a specific region with regard to oil spills. As an example, the ecologically very sensitive inner Jade Bight has been selected. A Lagrangian transport model was used to calculate particle trajectories which start in the vicinity of the location indicated by the ship symbol. For one decade, independent hypothetical releases have been simulated by initialising clouds of 250 particles once every 27 hours. For each of these releases the percentage of particles which enters the Jade Bight has been recorded. Oil weathering processes have not been considered so as to keep the results independent of any specific oil type, nor has the effect of countermeasures been taken into account. However, the simulations are already sufficient enough to enable the identification of examples of past weather conditions during which an oil spill would have led to particularly unfavourable impacts for the Jade Bight. More detailed analyses of hypothetical oil spill events under such realistic conditions may then be performed in a second step to assess and possibly improve oil spill contingency plans.

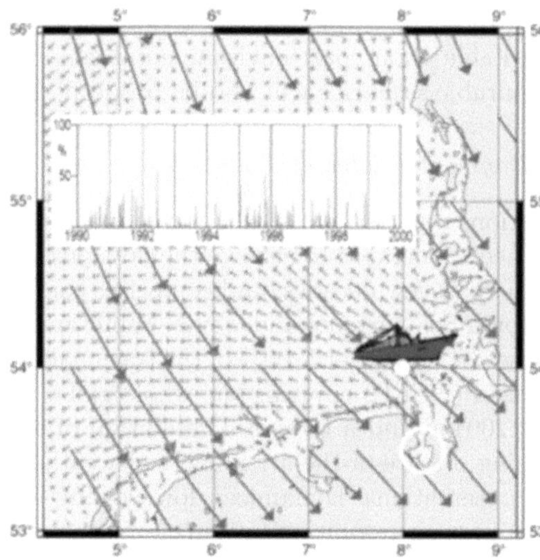

Fig. 7. Percentage of passive tracer particles released at the indicated location which would reach the particularly sensitive area of the Jade Bay (*white circle*). Fields of both currents and winds at an arbitrary time illustrate the type of information on which the analysis is based

An important constraint for the efficiency of countermeasures is the respite time until the oil slick actually hits the coast. The shorter the time window, the more dangerous an oil spill is. Again a large number of particle trans-

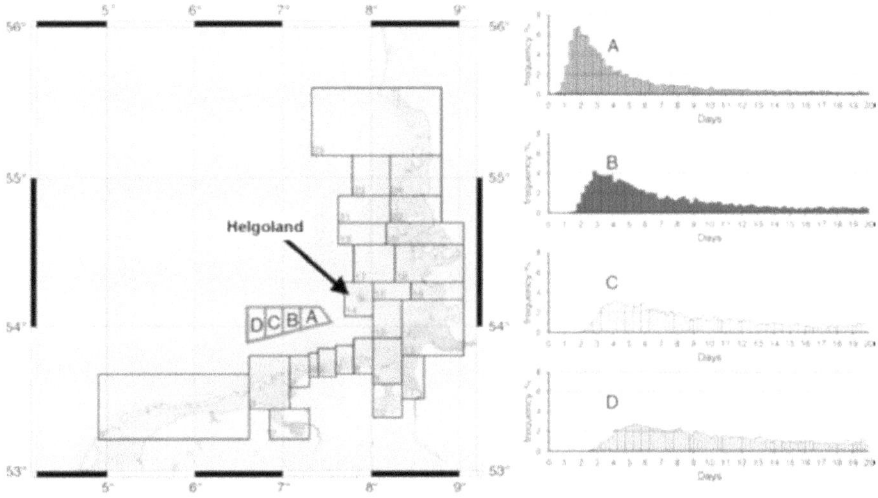

Fig. 8. Probability distributions of mean travel times from four assumed source regions (A–D) towards target box 14 centred on the island of Helgoland. The variable resolution unstructured grid over the North Sea is the basis for the hydrodynamic calculations

port simulations based on past weather conditions can be used to analyse the probability distribution of travel times, and in particular the probability of extremely short ones. Figure 8 depicts the results for particle clouds which are released from within the four regions A–D located next to the crowded shipping routes which connect the English Channel and the river Elbe. Box 14, containing the island of Helgoland, has been chosen as an example target region. It is noticeable that the width of the distribution of travel times increases with an increasing distance between source and target region.

A third example illustrates the possible use of the data set for the risk of ship colliding with off-shore wind parks. The example refers to a wind park assumed to be located close to the Dutch–German border in the vicinity to the main shipping routes (6°14' east longitude and 54°3' north latitude). The scenario underlying the analysis is that of a boat being disabled by a loss of command that starts drifting according to the prevailing winds and currents. Two multi-purpose ships ('Neuwerk' and 'Mellum') operating in the whole North Sea and one towing boat ('Oceanic') at a fixed stand-by location are supposed to be available to rescue the drifting ship. The positions of the two multi-purpose ships at the time of the accident are prescribed according to a random model. Considering a large number of realizations of the above scenario it is then possible to estimate the probability that a drifting ship will successfully be towed by one of the three rescue ships. A typical result is shown in Fig. 9. It should be emphasized, however, that modelling the probability

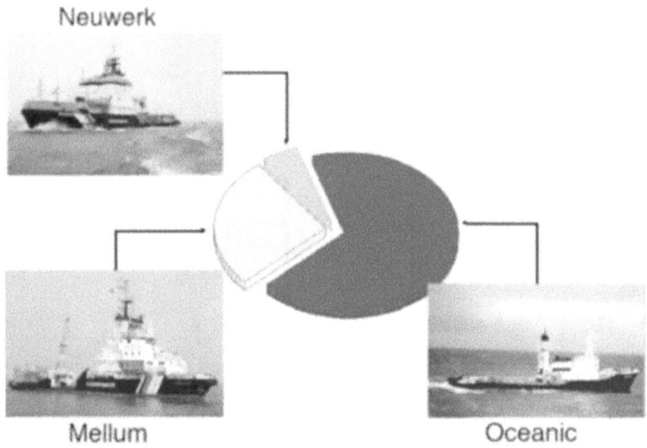

Fig. 9. Probability that a drifting ship will be towed by a specific towboat (Chitu, 2005)

of a successful rescue operation is difficult and depends considerably on the specific assumption without a sound statistical basis.

Acknowledgements

The fields of North Sea currents used in this study were produced by the Federal Waterways Research Institute, Coastal Division (BAW–AK) in Hamburg, Germany. Part of the work has been carried out in cooperation with the TU Delft. Contributions from Anca Hanea, Dana Chitu and Alin Chitu are gratefully acknowledged.

6 The Role of the European Maritime Safety Agency in Marine Pollution Response

Bernd Bluhm

On 24th November 2005, the European Maritime Safety Agency (EMSA) finalised contracts for the chartering of vessels for the provision of oil recovery services. Three companies have been selected following an open procurement process. A pool of five vessels will cover the Baltic Sea under the co-ordination of the Finnish company Lamor Corporation. The French company Louis Dreyfus Armateurs will provide one large vessel for the Atlantic Coast and western approaches to the English Channel and the Maltese company Tankship Management will operate in the Mediterranean Sea.

"This unique operational task is one of EMSAs major and most visible contributions to maritime safety in European seas" said Willem de Ruiter,

EMSA Executive Director. "By chartering these ships, EMSA fully participates in the protection of our seashores. The Prestige accident demonstrated the lack of high capacity response equipment in Europe and this weakness has now been partially remedied. The network of vessels will be further strengthened in 2006 and EU Member States will be able to rely on "a reserve for disasters" which will normally be available within a maximum of 24 hours".

The legal basis, approach and activities of the Agency in the field of pollution response are further detailed in this article.

Background

Following the accident of the oil tanker ERIKA in December 1999, the European Parliament and the Council adopted Regulation 1406/2002, which established the European Maritime Safety Agency (EMSA). In the aftermath of a new ecological catastrophe in European waters, caused in November 2002 by the accident of the oil tanker PRESTIGE, it became obvious that additional measures had to be taken on a pan-European level with regard to the response to ship-sourced oil pollution. Accordingly, the European institutions gave EMSA a new task in the field of oil pollution response through the publication of the amended Regulation (724/2004) on 29th April 2004.

The key objectives and tasks of the amended Regulation can be described as follows: EMSA is to facilitate an effective level of prevention of pollution and response to pollution by ships within the EU. The Agency has to provide Member States and the Commission with technical and scientific assistance in the field of accidental or deliberate pollution by ships. EMSA must, on request, support with additional means and in a cost-efficient way the pollution response mechanisms of Member States. More recently, the Agencys task to provide additional means in the field of monitoring and surveillance of marine pollution has been further elaborated. Specifically Article 10 of the Directive 2005/35/EC provides that EMSA is to assist the Member States in developing technical solutions and providing technical assistance for the tracing of discharges by satellite monitoring and surveillance.

Action Plan for Oil Pollution Preparedness & Response

In order to implement these tasks the Agency's Administrative Board, which is composed of representatives of each Member State of the EU, the European Commission, Norway and Iceland as well as representatives of industry, adopted the "Action Plan for Oil Pollution Preparedness and Response" in October 2004. This document is supplemented and updated by the EMSA Work Programme 2006. Both documents were developed in consultation with the Member States, the Commission and industry. Before determining the precise activities that the Agency should undertake to fulfil its legal obligations, it is necessary to outline some of the key points regarding the overall context for EMSA's activities.

Existing Framework. Having reviewed the approaches of Member States to response preparedness, it is clear that EMSA should also provide its support in the same spirit of co-operation and of supplementing resources and structures that are already in place. The OPRC 1990 Convention is the backbone of this attitude through its underlying tiered approach to spill response. Whilst it has been ratified by most Member States, there are distinct variations in the degree of implementation.

The Regional Agreements e.g. Helsinki Convention have made a significant contribution to improving preparedness and response to spills in Member States through the development of joint procedures and technical understanding of the issues.

Top-up Philosophy. As underlined by its Administrative Board, EMSA's operational task should be a "logical part" of the oil pollution response mechanism of coastal states requesting support and should "top-up" the efforts of coastal states by primarily focusing on spills beyond the national response capacity of individual Member States. EMSA should not undermine the prime responsibility of Member States for operational control of pollution incidents. The Agency should not replace existing capacities of coastal states. The Agency feels strongly that Member States have their own responsibilities regarding response to incidents. The requesting state will have the equipment at its disposal and under its sole command and control. The Agency's operational role should be conducted in a cost-efficient way. EMSA's activities should respect and build upon existing co-operation frameworks and regional agreements.

Action Plan: 3 Themes

The Agency undertook a review of various key issues concerning marine pollution in order to determine the high priority areas for the stationing of the additional at-sea oil recovery capacity. Issues reviewed included the historical incidence of spills in Europe. For example, Fig. 10 illustrates the location of tanker spills greater than 700 tonnes in Europe over the last 20 years. Those incidents involving more than 10,000 tonnes are highlighted in yellow. It is noteworthy that a significant proportion of these spills are in Western Europe.

Other factors reviewed included the present and future tanker trading patterns, the available (recognised) response options, an analysis of various case studies, the socio-economic and environmental sensitivities of the European coastline e.g. the Particularly Sensitive Sea Area (PSSA) of the Baltic Sea, as well as the activities of the Member States. The review included contributions from Member States, the Commission and the pollution response industry.

The Action Plan adopts a phase–in approach by firstly focusing on spills of heavy oil and with spills of other substances to be addressed at a later date. Consequently, the Action Plan identifies three main themes, namely:

- Operational Support
- Co-operation and Co-ordination

Fig. 10. Large tanker spills since 1984

- Information

Operational Support: At–sea Oil Recovery Services

Within the context of responding to large scale incidents involving heavy oils, the following technical points need to be noted: The most appropriate response strategy at the EU level for spills of heavy oil, or "weathered" oil, is by way of at-sea containment and recovery, Analysis of case studies indicate specific technical aspects that enhance at-sea recovery by vessels e.g. sweeping arm systems are generally more effective at recovering heavy oil than boom and skimmer systems,

There are various additional deficiencies in the response chain which should be addressed by all the parties concerned. These include the availability of appropriate monitoring and surveillance support, to assist the efficient deployment of anti-pollution vessels in the thicker concentrations of oil at sea.

Financial Framework

The budget for the main activity of 2005 i.e. the contracting of additional at-sea oil recovery capacity was € 17.5 m. The 2006 budget for at-sea oil recovery services is again € 17.5 m. It is evident that only limited resources are available to EMSA to carry out its Action Plan. Clearly the Agency does not have the financial resources available to buy or build dedicated oil pollution response vessels. With this in mind, the Agency has tried to offer at least a minimum viable system of additional means. In order to achieve this, a phasing-in period of some years is needed.

A European Public Private Initiative

Taking the previously mentioned issues into account, the Agency launched an open procurement procedure in order to work with the shipping and the spill response industry in providing at–sea oil recovery services to Europe. The Agency is providing the at–sea oil recovery capacity through 3 year contracts with the spill response and shipping industry. Such an approach has never been carried out before at the European level.

The vessel(s) under contract will, in normal circumstances, carry out its usual commercial activities. In the event of a large oil spill and following a request for assistance from a Member State, the vessel(s) will cease its usual activity and at short notice be transformed into and operate as an oil recovery vessel(s).

Appropriate modification or pre-fitting to the vessel(s) will be made in order to ensure that the specialised oil spill response equipment can be installed rapidly onboard the vessel and operated safely by the crew. Nonetheless it should be noted that all the vessels will be classed as at least "Occasional Oil Recovery" vessels.

The response capacity arrangements that have been contracted by the Agency are given in Table 5.

Under the EMSA arrangement, the contractors will offer at–sea oil recovery services from 31st March 2006 until 31st December 2008. The contracts may be renewed once. The pollution response equipment will be financed by EMSA under specific conditions.

It must be stressed that the vessels are a "reserve for disasters" at the disposal of every requesting coastal state to assist in cases of a (major) oil spill anywhere in European waters, the assistance provided by EMSA is not restricted to the indicated areas.

Each arrangement has the following common characteristics:

- The vessel will operate as an oil recovery vessel on the basis of a pre-agreed model contract with fixed fees and conditions as developed by the Agency for this purpose;
- The contractor is obliged to respond positively to all requests for assistance to respond to an oil spill, regardless of the spill location;

Table 5. At–sea Oil Recovery Contract Summary

At-sea Oil Recovery Services	Additional capacity (m^3)	Contract Value Value (€)	Lead Contractor
The Baltic Sea Arrangement	18,528 (max)	4,050,000	Lamor Corporation AB
The Atlantic and Channel Arrangement	4,000	8,500,000	Louis Dreyfus Armateurs SAS
The Mediterranean Arrangement	1,805–3,577	3,850,000	Tankship Management Ltd
Operational Fund*		2,280,000	
Overall Total		18,680,000	

*The Operational Fund will be used by EMSA to cover accelerated mobilisation of vessels immediately following a large scale spill and for the participation of vessels in multinational / regional / national at–sea oil recovery exercises.

- The primary oil recovery system is based around the "sweeping arm" concept (either rigid or flexible type) with an alternative "ocean going boom and skimmer" system also available. The requesting Member State can select the system in accordance with the incident characteristics;
- All the specialised oil spill response and associated equipment is containerised in order to facilitate rapid installation onboard the vessels;
- Each vessel has a speed over 12 knots for prompt arrival on site;
- Each vessel is equipped with a local radar based oil slick detection system;
- Each vessel has a high degree of manoeuvrability required to carry out oil recovery operations;
- Each vessel is able to decant excess water so maximising the utilisation of the onboard storage capacity;
- Each vessel has the ability to heat the recovered cargo and utilise high capacity screw pumps in order to facilitate the discharging of heavy viscous oils;
- Other complementary equipment comprises of flashpoint tester, oil/water interface system, gas detection (fixed and portable), sampling mini-lab and portable cleaning machines;
- Each crew will have been trained appropriately regarding the equipment and working under an international command and control structure. They will be able to provide the service on a 24 hour per day basis;
- Each vessel will be available for participation in at–sea spill response exercises e.g. BalEx Delta as undertaken within the framework of HelCom (maximum 10 days per year including sailing time).

2nd Round of Contracts in 2006 for At–sea Oil Recovery Services

The EMSA Work Programme 2006 confirms the "phasing-in" approach to build-up its "reserve for disasters" to an appropriate level. The Agency is organising a second call for tender in 2006 to build up the response capacity as planned. A Prior Information Notice was already published in the Official Journal of the European Union (OJEU) in January 2006. A third round will probably be launched for the Black Sea area after Romania and Bulgaria have become Member States of the European Union.

Satellite Monitoring and Surveillance Service

There is clearly a need for an operational system at the EU level for marine oil slick detection and monitoring as part of the response chain to locate accidental spills and illegal discharges. The added value at the European level will be:

- Achieving economies of scale – a combined purchase of satellite imagery should be more advantageous than on an individual national or regional basis,
- Providing continuity of service – the Agency would like to conclude a contract for 3 years,
- Improving the technical capabilities of the system – the Agency would like to discuss the technical modalities of satellite services with the providers to improve frequency, scope and timely availability of satellite data.

For such a large scale activity of the Agency, on the basis of Directive 2005/35/EC on ship-sourced pollution, a structure will be set-up in 2006 to support the activities of Member States and Commission in reacting to illegal discharges and accidental oil spills. EMSA is taking into account ongoing work in this field by the European Commission and others.

For this purpose, pre-analysed satellite imagery and information will most probably be purchased. Furthermore an appropriate infrastructure will be set up to provide advanced products (e.g. link with AIS, drift modelling) to complement the information for Member States, if requested. This will increase the reliability of the satellite imagery information and will be a value adding element presently missing in the national systems. The information from satellites should support the activities of Member States and the Commission primarily for the prevention of and response to illegal discharges from ships and, when required, for the recovery of oil from accidental spills.

Co-operation and Co-ordination

The European Community is contracting party to all the (major) regional agreements in Europe as illustrated in Fig. 11.

EMSA will upon request provide relevant Commission services with technical, operational and scientific assistance, i.e. to disseminate best practices among regional agreements. The regional bodies have expressed their interest in having close working relationships with EMSA in this particular field. In addition to the links with regional agreements, EMSA will work closely with the services of the European Commission within the existing co-operation mechanisms in an efficient way to avoid any duplication of activities.

Information

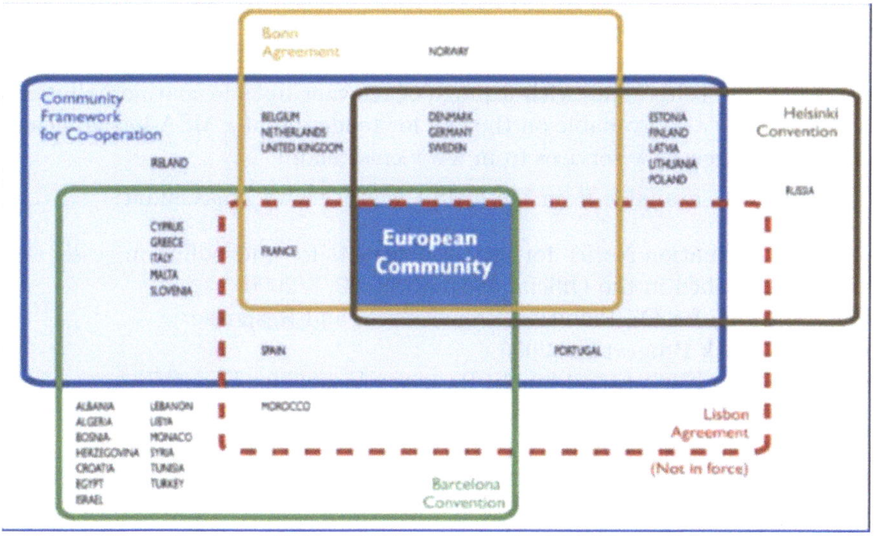

Fig. 11. International Framework for Co-operation in Combating Pollution

As previously stated, EMSA needs to provide the Commission and Member States with technical and scientific assistance. With the further development of the Agency's Pollution Response Unit, there will be capacity for the gathering, analysis and dissemination of best practices, techniques and innovation in the field of pollution response, in particular for at-sea oil recovery during large spills.

Experiences of recent accidents show that a close look is needed at the whole chain of activities required to be successful. In particular the issues of on-board oil storage and discharging recovered oil should be addressed. In previous incidents, vessels have spent a significant number of days in port discharging while they were needed at sea. Acceleration of this unloading phase is required.

It follows that, while the organisation of spill response is a key issue, it is essential that all the components required to conduct at-sea recovery be in place as the operation commences.

In a similar vein, many Member States are undertaking a review of the strengths and weaknesses of using chemical dispersants to response to marine oil spills. In this context EMSA has published an "Inventory of National Polices regarding the use of oil spill dispersants in the EU Member States" as well as a software tool to facilitate decision-making where the use of dispersants might be appropriate. Additional actions are also being undertaken in co-ordination with Member States.

An important activity is the dissemination of best practise to key parties. Accordingly, the Agency hosts a number of workshops or similar events. Speakers and delegates for such events come from a wide range of backgrounds including the spill response industry i.e. ITOPF (the International Tanker Owners Pollution Federation). The results of these events can be found on the Agency website along with a range of relevant links in marine pollution. Information is also available on the call for tenders that EMSA has launched for at-sea oil recovery services from www.emsa.eu.int.

Documents available from the EMSA website www.emsa.eu.int:

- Prior Information Notice for service contracts for anti-pollution vessel services published in the Official Journal the 20/1/2006
- Action Plan for Oil Pollution Preparedness and Response
- EMSA Work Programme 2006
- Information Note: Stand-by Oil Recovery Vessels for Europe
- Inventory of National Polices regarding the use of oil spill dispersants in the EU Member States
- Regulation (EC) N° 1406/2002 Establishing the European Maritime Safety Agency (EMSA)
- Regulation (EC) N° 724/2004 Amending Regulation (EC) N° 1406/2002

Conclusions

Most oil spills result from operational failures on all kinds of ships and oil platforms, not from major tanker accidents. The fate of oil at sea is determined by its characteristics and the physical, chemical and biological processes of the marine environment. Some lead to its dilution and removal, some cause it to change into persistent or more toxic forms. The timescale of these changes depends on the oil and on the weather and climatic conditions and they will affect the choice of clean-up techniques as well as the nature of the damage. Only a small percentage of oil will be biodegraded naturally in a reasonable amount of time.

Marine life can suffer from physical and chemical effects of the oil, especially in low energy environments with prolonged exposure e.g. in tidal flats. Long-lived species with low reproduction rates are most at risk. Sensitivity Maps are a valuable tool to plan response actions. They should be integrated into the overall oil spill contingency plans.

Advanced dispersants are able to control oil slicks. The "window of opportunity" can be kept open for a longer period. Today, some dispersants promise to break up and disperse even fuel oil and weathered oil. The use of dispersants is preferentially recommended in deep waters with high mixing energy and strong currents. The main advantage of dispersant application lies in the possibility of a rapid treatment of large areas by aircraft. Airborne assistance is essential for the guidance of ships and aircraft in oil-dispersant applications as well as for monitoring the success of the operation. The main consideration is to minimize the environmental harm. Decision-making tools to balance the pros and cons of dispersant application on a case-by-case basis are available. International agreements rule the assistance in case of severe oil spill incidents.

Although the multi-criteria ranking of options as one evaluation method used in Decision Support Systems is strongly affected by the choice of evaluation and weighting coefficients, uncertainty does not lower the power of Software Systems for enhanced Decision Making. DSS as well as integrated model systems, as introduced in the earlier chapters, provide a rational, albeit never perfect, base for a fast comparison of alternative countermeasures. In addition, they allow for testing containment strategies and oil spill plans in a systematic way. In comparison to standard and simple oil spill models which only forecast trajectories and the fundamental components of weathering, integrated model systems together with multilevel DSS enable a direct assessment of different response measures to be made, particularly with regard to potential hazards to the environment as well as to the economy. When implemented and used by regional and national authorities oil spill contingency planning and response systems provide enormous potential for enhancing the efficiency of decision making, even in politically sensitive situations.

Detailed reconstructions with a high resolution in space and time of past environmental conditions serve as a best estimate of natural conditions and allow a large number of Monte Carlo type experiments to be performed with realistic inputs. In particular, mutual interactions between winds, waves and currents are represented with sufficient detail. Conditioning simulations on past weather conditions at a given point in time allows for a consistent modelling of oil spill accidents and their environmental impact within a single framework. In this way the overall risk evaluation does not become biased by a missing correlation between unfavourable conditions for one or other aspect of the whole scenario. However, it must be kept in mind that the spectrum of scenarios which may occur is very wide and can never be studied or anticipated in all dimensions. Nonetheless, even if a risk study is unable to provide a reliable number for the overall risk, such an integrated study can still be useful in structuring the problem and making it more accessible for discussion between groups with vested interests.

The Action Plan for Oil Pollution Preparedness and Response (and the associated Work Programme 2006) has been put forward with the intention of strengthening the European response to oil pollution as requested by the Commission, the Council of Ministers and the European Parliament and, formally, through the regulation amending the tasks of EMSA. The Agency is in the process of implementing the identified tasks in co-operation with the Commission and the Member States. The most significant projects for 2006 are the new tenders for at–sea oil recovery services and the setting-up of the satellite imagery service.

The successful development of techniques and national/international agreements to increase the safety of navigation needs the co-operation of science, management and policy. The methods and strategies presented in this article describe some of the most recent and important tools in use for minimising the ecological and economical damage of oil spills.

Literature

Aamo OM, Reed M, Lewis A (1997) Regional Contingency Planning Using the OSCAR Oil Spill Contingency and Response Model. Proceedings of the 20th Arctic and Marine Oil Spill Program 1997, AMOP Technical Seminar, Vancouver

Baker, J.M. (1995): Net environmental benefit analysis for oil spill response. Proceedings, 1995 International Oil Spill Conference. American Petroleum Institute, Washington, D.C.: 611-614.

Belluck DA, Hull RN, Benjamin SL, French RD, OConnell RM (1993) Defining scientific procedural standards for Ecological Risk Assessment. In: J.W. Gorsuch, F.J. Dwyer, C.G. Iingersoll & T.W. LaPoint, eds. Environmental Toxicology and Risk Assessment. 2nd ed. American Society for Testing and Materials, Philadelphia, PA.: 440-450.

Bernem van, K.H. (1984): Behaviour of penetration and persistence of crude oil hydrocarbons in tidal flat sediments after artificial contamination. Senckenbergiana marit.; 16 (1/6): 13-30

Bernem van KH (2000) Conceptual models for ecology-related decisions. In: von Storch H, Flöser G (2000) Models in Environmental Research. Springer, Berlin, Heidelberg, New York

Bernem van, K.H., Drjes, J. and A. Mller (1989): Environmental oil sensitivity of the German North Sea coast. Proc. 1989 International Oil Spill Conference, American Petroleum Institute, Washington D.C., pp. 239-245, 1989

Bernem van, K.-H., Bluhm, B. and H. Krasemann (2000): Sensitivity mapping of particular sensitive areas, in: G.R. Rodriguez & C.A. Brebbia (eds.): Oil and Hydrocarbon Spills II; Modelling, Analysis and Control. WITpress, Southampton, Boston, ISBN 1-85312-828-7, pp. 229-238

Burns, K. (1993): Analytical methods used in oil spill studies. Mar. Poll. Bull., Vol. 26, No. 2: 68-72

Chitu, A. G., (2005): Probability of ship collision with offshore wind farms in the Southern North Sea, Master Thesis, Risk and Environmental Modelling, TU-Delft, 102 pp

Clark, R.B. (1992): Marine Pollution. Oxford University Press, Oxford

Clauss, G. (2003) Neue Entwicklungen seegangsunabhngige lskimmersysteme. Schiff & Hafen, Nr. 6, pp. 75-79

Dahling, P.S., Mackay, D., Johansen, O. and P.J. Brandvik (1990): Characterization of crude oil for environmental purposes. Oil and Chemical Pollution, v7: 199-24

European Topic Centre on Terrestrial Environment Topic Centre of European Environment Agency (2003): The Prestige Desaster. Available at http://terrestrial.eionet.eu.int/en_Prestige. Document last modified 2003/02/28

Feser, F., Weie, R. and H. von Storch, (2001): Multi-decadal atmopheric modeling for Europe yields multi-purpose data. EOS Transactions, 82, 305-310

Gilfilan, E.S. et al., (1983): Effects of spills of dispersed and non-dispersed oil on inter-tidal infaunal community structure. Proceedings 1983 Oil Spill Conference API Publication No. 4356, American Petroleum Institute, Washington DC, p. 457-463.

Gilfilan, E.S. et al., (1984): Effects of test spills of chemically dispersed and non-dispersed oil on the activity of aspartate amino-transferase and glucose-6-phosphate dehydrogenasein two intertidal bivalves, Mya arenaria and Mytilus edulis. Oil Spill Chemical Dispersants: Research, Experience and Recommendations, STP 840. T.E. Allen, Ed., American Society for Testing and Materials, Philadelphia, p. 299-313

Gundlach, E.R. and M.O. Hayes, (1978): Vulnerability of coastal environments to oil spill impacts. Mar. Tech. Soc. Jour. 12, 1827.

Gunkel, W. (1988): Ölverunreinigungen der Meere und Abbau der Kohlenwasserstoffe durch Mikroorganismen. In: Angewandte Mikrobiologie der Kohlenwasserstoffe in Industrie und Umwelt, R. Schweisfurth (ed), Kontakt & Studium Band 164, Technische Akademie Esslingen, Expert Verlag: 19-36.

IPIECA (2001): Dispersants and their role in oil spill response. IPICEA Report series Vol. 5, Nov. 2001.

Kistler R, Kalnay E, Collins W, Saha S, White G, Wollen J, Chelliah M, Ebisuzaki W, Kanamitsu M, Kousky V, van den Dol H, Jenne R, Fioriono M (2001) The NCEP/NCAR 50-year reanalysis: monthly means CD-ROM and documentation. Bull Am Met Soc 82:247–267

Leito, J.C., Laito, P., Braunschweig, F., Fernandes, R., Neves, R., Montero, P. (2003): Emergency activities support an operational forecast system – The Prestige accident. 4th Seminar of the Marine Environment 2003, Rio de Janeiro, Brasil

Lindstedt-Siva J (1991) US oil spill policy hampers response and hurts science. Proceedings, 1991 International Oil Spill Conference. American Petroleum Institute, Washington DC: 349–352

Liu X, Wirtz KW (2004) Using economic valuation methods to measure environmental damage in the coastal area. In: Proceedings of Littoral 2004. Aberdeen

Mansor SB, Assilzadeh H, Ibrahim HM, Mohamd AR (2002) Oil Spill Detection and Monitoring from Satellite Image. Map Asia Conference 2002, Bangkok

Mansor S, Pourvakhshouri SZ (2003) Oil Spill Management via Decision Support System. Risk Management FIG Regional Conference 2003, Marrakesch, Marokko

Meinke I, von Storch H, Feser F (2004) A validation of the cloud parameterization in the regional model SN-REMO. J Geophys Res 109, D13205, DOI:10.1029/2004JD004520

Michel J, Dahlin J (1993) Guidelines for Developing Digital Environmental Sensitivity Index Atlases and Databases. National Oceanic and Atmospheric Administration, Seattle

Operational Response Guidelines (2003):. Ministry of Transport, Public Works and Watermanagement, Rijswijk, NL, 2003

O'Sullivan, A. J. and T. G. Jacques (2001): Community Information System for the Control and Reduction of Pollution, IMPACT REFERENCE SYSTEM, Effects of Oil in the Marine Environment: Impact of Hydrocarbons on Fauna and Flora. Version 2001

Page, D.S. et al., (1993): Long-term fate of dispersed and undispersed crude oil in two nearshore test spills. Proceedings 1983 Oil Spill Conference. API Publication No. 4356, American Petroleum Institute, Washington DC, p. 465-471. Response to Marine Oil Spills. The International Tanker Owners Pollution Federation Ltd 1993

Sea Empress Environmental Evaluation Committee (SEEEG) (1996): Initial Report, July 1996

Serret, P., Álvarez-Salgado, X.A., Bode, A. and 419 other scientists from 32 universities and 6 research institutions (2003): Spain's earth scientists and the oil spill. Science-Magazine, Vol. 299, Letters to the Editor

The Sea Empress incident (1996):. Report by the Marine Pollution Control Unit. The Coast Guard Agency, Southampton, 1996

Sensitivity Mapping for Oil Spill Response (1994): IMO/IPIECA Report Series. Volume One 1994

The Halifax Declaration on The Ocean (1998): PACEM IN MARIBUS XXVI, 29 November – 3 December, 1998: Halifax, Canada

UNCTAD (1998): Review of Maritime Transport 1998, Geneva, 1998

US ENVIRONMENTAL PROTECTION AGENCY (1992a) Framework for Ecological Risk Assessment. US EPA Document: EPA/630/R-92/001. US Environmental Protection Agency, Washington DC

Using dispersants to treat oil slicks at sea (2005): Aiborne and shipborne treatment. Response manual. Cedre, Brest, December 2005.

Weie R, PlüßA (2005) Storm-related sea level variations along the North Sea coast as simulated by a high-resolution model 1958–2002. Ocean Dynamics, in print

Wirtz KW, Adam S, Liu X, Baumberger N (2004) Robustness against different implementation of uncertainty in a contingency DSS: The Prestige oil spill revised. International Environmental Modelling and Software Society iEMSs 2004 International Conference, Osnabrück

Wirtz KW, Baumberger N, Adam S, Liu X (2005) Oil spill impact minimization under uncertainty: evaluating contingency simulations of the Prestige accident. Submitted to Ecological Economics

Zigic S (2004) Ocean Modelling – Systems Available for the Environment and Scientific Coordinator. Environment and Scientific Coordinators (ESC) Workshop 2004, Tasmania, Australia

Index

MIX
Papier aus verantwortungsvollen Quellen
Paper from responsible sources
FSC® C105338
FSC
www.fsc.org

If you have any concerns about our products,
you can contact us on
ProductSafety@springernature.com

In case Publisher is established outside the EU,
the EU authorized representative is:
Springer Nature Customer Service Center GmbH
Europaplatz 3, 69115 Heidelberg, Germany

Printed by Libri Plureos GmbH
in Hamburg, Germany